DEVELOPING EDITING SKILL

Sue C. Camp

Assistant Professor
Broyhill School of Management
Gardner-Webb College
Boiling Springs, North Carolina

Gregg Division/McGraw-Hill Book Company

New York   Atlanta   Dallas   St. Louis   San Francisco   Auckland
Bogotá   Guatemala   Hamburg   Johannesburg   Lisbon   London
Madrid   Mexico   Montreal   New Delhi   Panama   Paris
San Juan   São Paulo   Singapore   Sydney   Tokyo   Toronto

Sponsoring Editor / Roberta Moore
Editing Supervisor / Sharon E. Kaufman
Design and Art Supervisor / Frances Conte Saracco
Production Supervisor / Priscilla Taguer

Text Designer / Delgado Design, Inc.
Cover Designer / The Design Source

DEVELOPING EDITING SKILL

567890 SEMSEM 8910987

ISBN 0-07-009638-4

Technological changes are taking place in today's business environment, making it more important than ever that office workers have the skills employers say they need the most. Employers put the ability to communicate well at the top of their needs list. This means not only being able to speak and listen effectively and to understand what you read—basic skills that are absolutely essential. It also means being able to produce written communications that are error-free and that represent the highest professional standards. In order to meet this requirement you will need to become competent at editing business communications.

You may have already acquired skill in proofreading—that is, checking final copies of documents to make sure they are free of errors. *Developing Editing Skill* will help you acquire the skills you need to become competent in editing as well. Editing is revising a communication, often while it is still in rough draft form. Obviously, this includes correcting errors in spelling, grammar, and punctuation. However, editing is much more than just correcting mechanical errors.

Editing requires looking at a document critically to see if it can be improved in a variety of ways and answering six major questions: *IS IT CORRECT?* Are there errors in punctuation, grammar, or spelling? Is the information accurate? *IS IT COMPLETE?* Have necessary details been omitted? Does the style and format conform to accepted practice? *IS IT CLEAR?* Have words been used effectively to get across the writer's intended meaning? *IS IT COURTEOUS?* Has the document been written with the reader's viewpoint in mind? *IS IT CONCISE?* Has the writer used too many words? Are sentences and paragraphs kept to a length that the reader can easily manage? *IS IT CONSISTENT?* Are there conflicting facts in the document? Are similar items treated the same? Is usage consistent? Is information presented in logical sequence?

Once you have made certain that all the questions above have been resolved and the document has been revised—edited and proofread—until it is as close to perfect as possible you are then ready to produce *THE FINISHED PRODUCT. Developing Editing Skill* is designed to improve your competence in all of these areas.

Here are some special features that you will find in *Developing Editing Skill:*

1. A variety of applications covering business documents that are likely to be encountered on the job.
2. Emphasis on how to solve the most frequently occurring problems and apply the rules that are most often abused.
3. Special revision symbols that can be applied to word processing functions.
4. A reference section that will help you develop the "look-it-up" habit.
5. Two kinds of review exercises at the end of each chapter. Exercise A reviews applications learned in the current chapter. Exercise B reviews applications in the current chapter and previous chapters.

Whether you plan to start work soon or you are already on the job, whether you work in a traditional office or an electronic one, *Developing Editing Skill* will help you become competent at producing effective written communications.

Sue C. Camp

CONTENTS

WHAT IS EDITING?

CHAPTER

Editing Skill

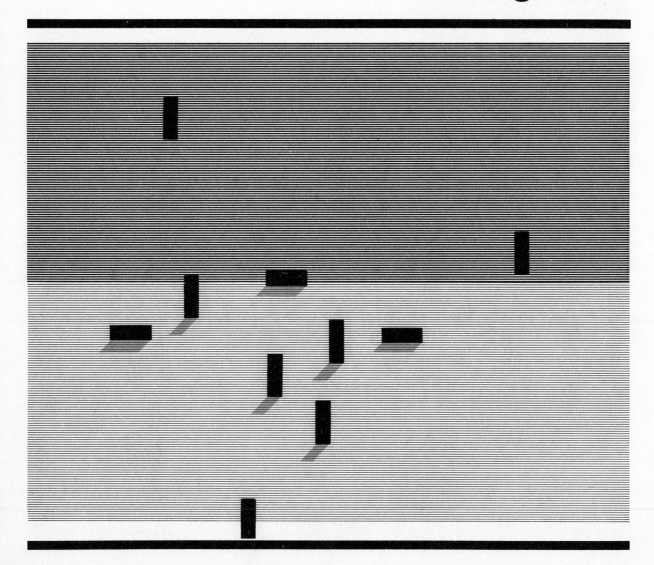

What Is Editing?

As you begin reading this book, you are probably pondering the question, "What is editing?" **Editing** is revising a written communication to improve it.

Editing is a very important skill that is made up of other skills. You may need to review some of these skills, such as how to punctuate correctly (editing for correctness). Other skills, such as logical sequencing of information (editing for consistency), may be new to you.

In editing, you assume the role of business communications "doctor." You diagnose the problem and prescribe the treatment. The prescription is in the form of revision symbols.

Let's begin editing with a skill that may be familiar to you, marking for capitalization.

CAPITALIZATION SYMBOLS

Capitalization revision symbols are shown in the following table. The column, which is entitled "Revision," explains the change that is to be made. The second column, "Edited Draft," shows how to actually mark the change on a draft. "Final Copy," which is the last column, shows how the corrected copy will appear. Study the revision symbols before doing Exercise 1-1.

CAPITALIZATION

Revision	Edited Draft	Final Copy
Capitalize a letter	texas	Texas
Lowercase a letter	This	this
All capitals	Cobol	COBOL
Lowercase a word	PROGRAM	program
Initial capital only	PROGRAM	Program

EXERCISE 1-1

Using appropriate revision symbols, mark the changes. Then write or type the corrected sentence.

1. Use all capitals for *Cobol.*

   Does your programmer know Cobol?

   Does your programmer know COBOL?

2. Use no capitals in *PROGRAM.*

   She wrote our inventory control PROGRAM.

   She wrote our inventory control program.

NAME _____   DATE _____

3. Use an initial capital in *thursday*.

   The convention begins thursday.

   The convention begins Thursday.

4. Use initial capital only in *APRIL*.

   Will you have the figures by APRIL 1?

   Will you have the figures by April 1,?

5. Do not capitalize *Leave*.

   When do you plan to Leave?

   When do you plan to leave?

Word Processing Terminology

People who write with and edit on word processing equipment need to be familiar with some word processing words and terms.

A **document** is a written, typed, keyboarded, or otherwise produced, communication. A document may be on paper, or it may be stored in a word processor or a computer. Most documents can be improved in many ways.

The **originator** is the person who initiates a document. Originators may dictate or handwrite a message, or they may keyboard it.

Keyboarding is typing a document into a standalone word processor or into a computer that has word processing capabilities.

A **standalone word processor** is a self-contained (not linked with a computer) unit with text manipulation capabilities. Its primary function is word processing.

A **microcomputer** is a small computer—often not much bigger than a typewriter. Many small and large computers have word processing capabilities that allow communications to be keyboarded, changed, corrected, manipulated, and printed.

Word processing capabilities can either be part of the computer's permanent, internal memory or the function of software. **Software** is an externally stored program written for a specific purpose. With word processing software, computers can perform many of the same functions performed by a standalone word processor.

How Does Editing Relate to Word Processing?

The user of word processing equipment ordinarily enters documents by keyboard and views them on a screen called a cathode-ray tube (CRT). Changes are made—for example, words and paragraphs are moved around—with the touch of a button. Most of the equipment is **user friendly,** which means that it can be operated without extensive training.

Executives who have access to word processors or computers with word processing capabilities may choose to keyboard their rough drafts for several reasons:

1. Keyboarding is a quick way to put thoughts into written form.
2. Keyboarding is convenient because it can be done after usual business hours.
3. With keyboarding, only one person's time is required in originating a document.

After documents are keyboarded in rough draft, they must be edited into mailable form. For example, it may be necessary to insert or delete letters,

NAME ━━━━━━━━━━━━━━━━━━━━━━━━━━━━ DATE ━━━━━━

spaces, words, or punctuation, or to make other kinds of changes and corrections. Editing can be done before the document is printed. Frequently, however, the document is printed and given to the originator for editing. The originator marks the changes, and the document is returned to the word processing operator who makes the changes.

Using Revision Symbols

Whether you are editing handwritten drafts, shorthand notes, typed drafts, or messages viewed on a screen, the editing process is the same. Errors and other weaknesses must be found and corrected. You have already seen how capitalization changes are marked. Learn how other revisions are indicated by studying the following charts and completing the related exercises.

CHANGES AND TRANSPOSITIONS

Revision	Edited Draft	Final Copy
Change a word	price is only $10.98 . . . *$12.99*	price is only $12.99 . . .
Change a letter	deduct*i*ble	deductible
Stet (do not make the change)	price *are* is only $10.98 . . .	price is only $10.98 . . .
Spell out	on Washburn (Rd.) in . . .	on Washburn Road in . . .
	(2) pens and (4) pencils . . .	two pens and four pencils . . .
Move as shown	. . . on May 1 write him . . .	. . . write him on May 1 . . .
Transpose letters	hte	the
Transpose words	most the of staff	most of the staff

EXERCISE 1-2

Using appropriate revision symbols, mark the changes. Then write or type the corrected sentences.

1. Change price from *$18.98* to *$17.98*.

   The invoice was only $18.98.

   The invoice was only $ 17.98.

2. Ignore the change.

   What is the ~~date~~ *day* of the letter?

   What is the date of the letter?

3. First, mark the transposition. Then put *and memos* after *letters*.

   We received letters form him and memos.

   We receive letters and memos from him

NAME ▬▬▬▬▬▬▬▬▬▬▬▬▬▬▬▬▬▬▬▬▬▬▬▬▬ DATE ▬▬▬▬▬▬▬▬

4. Mark the transposition.

All us of agreed.

All of us agreed.

5. Spell out the number. Spell out the abbreviation.

Tom will buy 2 tickets by Mon. night.

Tom will buy two tickets by Monday night.

INSERTIONS

Revision	Edited Draft	Final Copy
Insert a letter	may leave erly	may leave early
Insert a word	in the office	in the office
Insert a comma	may leave early . . .	may leave early, . . .
Insert a period	Dr Maria Rodriguez	Dr. Maria Rodriguez
Insert an apostrophe	all the boys hats	all the boys' hats
Insert quotation marks	Move on, she said.	"Move on," she said.
Insert hyphens	up to date report	up-to-date report
Insert a dash	They were surprised even shocked!	They were surprised—even shocked!
Insert parentheses	pay fifty dollars ($50)	pay fifty dollars ($50)
Insert one space	mayleave	may leave
Insert two spaces*	1.The new machine	1. The new machine

* Use marginal notes for clarification.

DELETIONS

Revision	Edited Draft	Final Copy
Delete a letter and close up	strooke or strooke	stroke
Delete a word	wrote two ~~two~~ checks	wrote two checks
Delete punctuation	report was up to date.	report was up to date.
Delete one space*	good day	good day
Delete space	see ing	seeing

* Use marginal notes for clarification.

NAME _____   DATE _____

EXERCISE 1-3

Mark the indicated changes using appropriate revision symbols. Then write or type the corrected sentence.

1. Insert the word *your* after the word *is*.

   What is social security number?

   What is your social security number?

2. Insert the *n* in *inventory*.

   Your invetory should be insured.

   Your inventory should be insured.

3. Insert a comma between the city and the state.

   Our branch is in Lincoln Nebraska.

   Our branch is in Lincoln, Nebraska.

4. Insert an apostrophe before the *s* in *Lindas*.

   Karen Solini asked for Lindas address.

   Karen Solini asked for Linda's address.

5. Insert the needed spaces.

   What is the price?Has it changed?

   What is the price? Has it changed?

EXERCISE 1-4

Mark the changes using the appropriate revision symbols. Then write or type the corrected sentences.

1. Delete unneeded stroke in *report*.

   Type the repoort as soon as possible.

   Type the report as soon as possible.

2. Delete the second *the*.

   Please complete the the report before our meeting.

   Please complete the report before our meeting.

3. Delete the comma.

   He talked with Jan, and Tom.

   He talked with Jan and Tom.

4. Delete the extra space between words.

   Good managers delegate responsibilities.

   Good managers delegate responsibilities.

5. Close up the space after the comma.

   We have 1, 206 cases in inventory.

   We have 1,206 cases in inventory.

What Is the Impact of Office Technology?

Editing skill has always been important, but advances in office technology have caused renewed emphasis. Word processing equipment and computers have made editing easier. For years, business people have processed words through dictation and transcription, but office automation has simplified this process. Editing skill is important for office professionals, such as secretaries, administrative assistants, word processing personnel, and managers, who are involved with written communications.

Editing skill is essential in deciding *when* to use *which* machine function to change *what* to *what* or to move something from *here* to *there*. The equipment can perform only the functions that the operator tells it to perform. The human decision-making ability is essential in efficient editing. Without editing skill, expensive, automated equipment can't do much to improve business communications.

What Is the Goal of This Book?

The goal of this book is to help you improve your written communication skill through editing. You can achieve this goal by becoming aware of weaknesses that occur frequently in business writing and learning how to find these weaknesses and correct them.

How are weaknesses found and corrected? Weaknesses are found and corrected through competent editing. Competent editors of business writing look for the qualities in the Editing Checklist that follows. Components of these broad classifications will be examined in the chapters that follow.

Editing Checklist
- ✔ Correct?
- ✔ Complete?
- ✔ Clear?
- ✔ Courteous?
- ✔ Concise?
- ✔ Consistent?

REVIEW EXERCISE 1-A

Part A

Circle T *for* true *and* F *for* false.

1. T F   Editing is an important skill made up of other skills.
2. T (F)   Documents must be printed on paper.
3. (T) F   The person who initiates a document is called the originator.
4. (T) F   Typing a document into a word processor is called keyboarding.
5. T (F)   A standalone word processor is always linked with a computer.
6. T (F)   Executives frequently keyboard their rough drafts because it is a quick, convenient way to originate a document.

NAME ▬▬▬▬▬▬▬▬▬▬▬▬▬▬▬▬▬▬▬▬▬ DATE ▬▬▬▬

7. T F Even with expensive, automated equipment, operators need to develop editing skill.

8. T (F) Editing does *not* involve decision making.

9. T (F) Editing is a mental function—not just a machine function.

10. (T) F Weaknesses in communications are found through competent editing.

Part B

Circle the letter of the best answer.

1. Software is an externally stored program that is
   (a.) Written for a specific purpose.
   b. Often used to add word processing capabilities.
   c. Not part of the computer's memory.
   d. All of the above.

2. Editing skill is needed by
   a. Managers—not secretaries.
   b. Secretaries—not managers.
   (c.) All office professionals involved with written communication.
   d. Administrative assistants and word processing supervisors only.

3. A CRT is a
   a. Control-ray typewriter.
   (b.) Cathode-ray tube.
   c. Cathode-ray turntable.
   d. Control-ray tube.

4. Word processing equipment
   a. Is designed to shred confidential papers.
   b. Corrects all mistakes automatically.
   (c.) Has made editing much easier.
   d. Usually doesn't have editing capabilities.

5. Editing can be done
   a. With pen or pencil on a typed or handwritten draft.
   b. While viewing the message on the screen.
   c. On shorthand notes that haven't been transcribed.
   (d.) All of the above.

6. Computers with programs that can be operated without extensive training are
   (a.) User friendly.
   b. Computer-operator noted.
   c. User-file manufactured.
   d. Distributor activated.

NAME ▬▬▬▬▬▬▬▬▬▬▬▬▬▬▬▬▬▬ DATE ▬▬▬▬▬

7. Revising a communication to improve it is called
    a. Filing.
    b. Editing.
    c. Distributing.
    d. Coding.

8. Efficient editors should determine if the communication is
    a. Correct, complete, and clear.
    b. Courteous, concise, and consistent.
    c. Correct answer not given.
    d. Answers *a* and *b*.

9. Automated office equipment
    a. Cures all communication problems.
    b. Does not require an operator.
    c. Is effective only when used by competent personnel.
    d. Makes editing more difficult.

10. The goal of this book is to help you improve written communications through editing by increasing your
    a. Awareness of frequent weaknesses.
    b. Knowledge of how to find weaknesses.
    c. Knowledge of how to correct weaknesses.
    d. All of the above.

REVIEW EXERCISE 1-B

Use appropriate revision symbols to make these corrections on the memorandum draft, which appears on page 10.

1. Insert a colon after *To*.
2. Use initial capital only in *FREIDA*.
3. In the heading, use initial capital only for *FROM*.
4. Insert the correct year—*Date: April 30, 19—*.
5. Delete the second *p* in *shippment*.
6. Transpose *to* and *us* to read *us to*.
7. Add two spaces after the period ending the first sentence.
8. Change the shipping date to *May 4*.
9. Change *ie* to *ei* in *Frieght*.
10. Spell out *2*.
11. *Trucks* shouldn't be capitalized.
12. Delete the second *call* in the last paragraph.
13. Close up space in *de layed*.
14. Insert *any* between *for* and *reason*.

NAME ━━━━━━━━━━━━━━━━━━━━━━━━━━ **DATE** ━━━━━━

MEMORANDUM

To   FREIDA Mason, Production Scheduling
        Department
FROM: Luke Edwards, Shipping Department
Date: April 30, 19
Subject: Fielding Company Shippment

Fielding asked to us ship their order as soon as
possible. It should be finished and shipped May 5. 4

Fast Frieght Carriers will send 2 Trucks May 5, at
3:30 p.m. Call call me if this order is de layed
for reason.

kd

MEMORANDUM

To: Freida Mason, Production Scheduling
       Department.
From: Luke Edwards, Shipping Department
Date: April 30, 1990
Subject: Fielding Company Shipment

Fielding asked us to ship their orders as soon
as possible. It should be finished and shipped May 4,

Fast Freight Carriers will send two trucks MAY 5,
at 3:30 p.m. Call me if this order is delayed for
any reason.

kd

NAME _____ DATE _____

CHAPTER

2

Revision Symbols

Correct Messages Are Essential

Correct messages are a vital part of successful business operations. Written business messages are frequently the first contact a potential client has with a company. First impressions made by this initial contact are important in establishing a new relationship. A potential client receiving carelessly written or incorrect messages may assume that business transactions will be handled in the same way. Editing helps project a quality image through effective business communications.

Frequently, the quality of written communications goes unnoticed until there is a problem resulting from a poorly written or incorrect communication. Even though the problem may be smoothed out, the damage has been done. Had appropriate editing techniques been used, the problem would have been prevented.

Revision Symbols Used in Editing

Producing a good quality communication from an edited draft requires an efficient method for marking needed changes. Needed changes must be indicated for the typist in a quick, easy-to-follow way. Revision symbols, which are quick and easy-to-follow, can reduce turnaround time. **Turnaround time** is the processing time needed to complete and return the document to the originator. Using revision symbols helps the typist make the intended changes correctly.

In Chapter 1, you learned the symbols for capitalization, changes, insertions, deletions, and transpositions. In Chapter 2, you will learn block, global, and formatting revision symbols.

BLOCK SYMBOLS

In editing a communication, you frequently must add, delete, change, or move certain segments. These segments are usually called **blocks.** Word processing equipment has made it possible to manipulate (change in some way) these blocks of text without rekeyboarding the entire document.

Sometimes large blocks (such as paragraphs or pages) are manipulated. Other times, small blocks (only two or three words) are manipulated. To avoid confusion, use letters to identify blocks.

Identified Block

Ralph /Ạnd Glenda/ planned the menu.

After identifying blocks, you can insert, delete, move, and query them.

A query is a question mark in the margin opposite a block of information that must be verified. You may clarify the query by asking the document originator or by checking an appropriate, reliable source. Here are some examples.

1. Query blocks that may be either right or wrong.

Ed Bates will retire in May at the age of 96. ? ⒝

NAME ━━━━━━━━━━━━━━━━━━━━━━━━━━━━━━━━ DATE ━━━━━━

Is 96 right, or were the numbers transposed? Find out!

2. Query blocks that definitely have an error.

   Make my plane reservation for June 31. ? C

June has only 30 days. Should the reservation be for June 30 or July 1? Find out!

3. Query blocks with contradicting information.

   I will call Monday morning/at 8:30 p.m.   D ? E

Is it *morning* or *p.m.*? Find out!

BLOCK SYMBOLS

Revision	Edited Draft	Final Copy
Identify block	/and the catalog/will be mailed.	
Insert identified block	Your order␣will be mailed.	Your order and the catalog will be mailed.
Delete identified block	Your order/and the catalog/will be mailed.	Your order will be mailed.
Move identified block*	Your order/and the catalog/will be shipped. The invoice␣will be mailed. *move* A	Your order will be shipped. The invoice and the catalog will be mailed.
Query identified block*	Ed will retire at the age of 96. ? B (Are the numbers transposed? Verify age.)	Ed will retire at the age of 69.
Query identified block*	Make my reservation for June 31. ? C (June has only 30 days. Verify date.)	Make my reservation for June 30.
Query conflicting blocks*	Call me Monday/morning/at 8 D ?E /p.m./ (*Morning* or *p.m.*? Verify time.)	Call me Monday morning at 8 a.m.

* Use marginal notes for clarification.

EXERCISE 2-1

Complete each item as indicated.

1. Identify block A. Block A: *of the Houston store*

   . . . promotion to manager /of the Houston store/ last week . . .

2. Identify block B. Block B: *suggested telemarketing*

   . . . sales manager /suggested telemarketing/ to increase . . .

NAME ━━━━━━━━━━━━━━━━━━━ **DATE** ━━━━━━━

3. Identify and query conflicting blocks. Block E: *twelve memos* Block F: *five before and five*

He dictated twelve memos—five before and five after lunch. ⬛E⬛ ? ⬛F⬛

4. Identify and delete block C; then write the new sentence. Block C: *and memos*

The letters and memos were mailed on December 1.

The letters were mailed on December 1.

5. Mark block D and insert Block D after *letters*. Then write both new sentences. Block D: *and the catalog*

The samples and the catalog arrived today.

The letters arrived yesterday.

The samples arrived today.

The letters and the catalog arrived yesterday.

GLOBAL SYMBOLS

On most word processing equipment, you can make identical inserts, deletions, or changes throughout the document by using the global function. The **global function** instructs the machine to alter the document each time the incorrect text occurs. When you are editing a document prepared on equipment with global functions, you need mark only the first occurrence of the incorrect text. Circle a symbol in the margin to indicate a global change.

Study the global symbols below before completing Exercise 2-2.

GLOBAL SYMBOLS

Revision	Edited Draft		Final Copy
Global insert* (Inserts *red* with every occurrence of *labels*)	all ^labels. The large labels are . . . (red labels)	⌃	all red labels. The large red labels are . . .
Global delete* (Deletes every occurrence of *very*)	a ~~very~~ high profit on a very good item	⟨⟩	a high profit on a good item
Global change* (Changes every occurrence of *very high* to *high*)	a ~~very high~~ profit on a very good item. Our very high profit . . . (high)	⊖	a high profit on a very good item. Our high profit . . .

* Use marginal notes for clarification. Marginal symbols are circled to indicate changes are global.

EXERCISE 2-2

Assuming that you have global functions on your equipment, mark the changes and write the new sentences. Remember, mark only the first occurrence of the needed change.

1. Global insert: *free* before each occurrence of *catalog*.

Your advertisement offered a ^catalog. Please send the catalog by . . . (free catalog)

Your advertisement offered a free catalog. Please send the free catalog by . . .

2. Global delete: *very* before each occurrence of *high*.

Sales are ~~very~~ high, but profits are very low. Costs are very high . . .

Sales are high, but profits are very low. Costs are high . . .

3. Global change: *1,230* to *1,320* each occurrence.

Inventory figures show ~~1,230~~ cases. All 1,230 cases will be . . .

Inventory figures show 1,320 cases. All 1,320 cases will be . . .

4. Global delete: *and typists* each occurrence.

All secretaries and typists were invited to the seminar. Only ten secretaries and typists attended.

All secretaries were invited to the seminar. Only ten secretaries attended.

5. Global change: *page 32* to *page 35*.

The promotion policy is on ~~page 32~~. Please read page 32 carefully.

The promotion policy is on page 35. Please read page 35 carefully.

FORMAT SYMBOLS

Format symbols indicate that certain words in a document should be treated differently from the way the rest of the copy is treated. For example, a document may contain headings, titles, or other items that should stand out in order to aid the reader in finding information, to indicate special emphasis, or to conform to accepted style.

Another use for format symbols is to indicate how a document or parts of a document should be placed on a page so that the page is visually appealing and conforms to accepted style. If you are working on a document that contains many different elements, it might be necessary to type or keyboard a draft. If you were keyboarding on a word processor, you would then print a copy of the draft and use revision symbols to show the changes that need to be made so that the document is set up properly.

Study the following format symbols for boldface, underscore, paging, paragraphing, centering, alignment, and spacing.

FORMAT SYMBOLS: BOLDFACE AND UNDERSCORE

Revision	Edited Draft	Final Copy
Print boldface	Bulletin	**Bulletin**
No boldface	**Bulletin**	Bulletin
Underscore	Title	Title
No underscore	Title	Title

NAME ▬▬▬▬▬▬▬▬▬▬▬▬▬▬▬▬ **DATE** ▬▬▬▬▬▬

EXERCISE 2-3

Mark the indicated changes with the correct revision symbol.

1. Print boldface.

   Current Assets

2. No boldface.

   Cash and Marketable Securities

3. Underscore.

   INCOME STATEMENT

4. No underscore

   <u>Current Liabilities</u>

5. Underscore and print boldface.

   INCOME STATEMENT

FORMAT SYMBOLS: PAGE AND PARAGRAPH

Revision	Edited Draft	Final Copy
Begin a new page	. . . order was delivered today by *pg.* common carrier. We have all the materials for the product.	. . . order was delivered today by Page 2 common carrier. We have all the materials for the product.
Begin new paragraph	. . . order was delivered today by common carrier. ¶ We have all the materials for the product. It should . . .	. . . order was delivered today by common carrier. We have all the materials for the product. It should . . .
No new paragraph (run-in)	. . . order was delivered today by common carrier. *No* ¶ We have all the materials for the product. It should . . .	. . . order was delivered today by common carrier. We have all the materials for the product. It should . . .
Indent five spaces	*5* We have the raw materials in our warehouse. Production will . . .	We have the raw materials in our warehouse. Production will . . .

EXERCISE 2-4

Write the symbols for each of the following items.

1. Begin a new paragraph.

   _____

2. Begin a new page.

   _____

NAME _____ DATE _____

3. No new paragraph.

    No ¶

4. Indent five spaces.

    5

FORMAT SYMBOLS: CENTERING

Revision	Edited Draft	Final Copy
Center line horizontally	⌐TITLE⌐	TITLE
Center identified block horizontally and vertically*	A MENU Juice and Coffee Scrambled Eggs Toast and Jam	MENU Juice and Coffee Scrambled Eggs Toast and Jam

* Use marginal notes for clarification.

FORMAT SYMBOLS: ALIGNMENT

Revision	Edited Draft	Final Copy
Align horizontally	Coleen answered the letter.	Coleen answered the letter.
Align vertically	Coleen answered the letter today. You should receive it Monday. $122.30 22.40 $144.70	Coleen answered the letter today. You should receive it Monday. $122.30 22.40 $144.70

FORMAT SYMBOLS: SPACING

Revision	Edited Draft	Final Copy
Single-space	ss ⌐ xxxxxxxxxx xxxxxxxxxx	xxxxxxxxxx xxxxxxxxxx
Double-space	ds ⌐ xxxxxxxxxx xxxxxxxxxx	xxxxxxxxxx xxxxxxxxxx
Triple-space	ts ⌐ xxxxxxxxxx xxxxxxxxxx	xxxxxxxxxx xxxxxxxxxx

NAME ▬▬▬▬▬▬▬▬▬▬▬▬ **DATE** ▬▬▬▬▬

EXERCISE 2-5

Using the appropriate revision symbol, mark the indicated changes.

1. Center the line horizontally.

   ⌐ JESSICA C. ABRIZONI, CPA ⌐

2. Identify the entire block and then center it vertically and horizontally.

   JESSICA C. ABRIZONI, CPA

   Announces a New Location

   One Plaza Tower

   Shelby, North Carolina 28150

3. Single-space this copy.

   Mr. Steve Camden

   ss Post Office Box 83

   Columbia, SC 29602

4. Double space this copy.

   ds   Reports are double-spaced
   for many reasons . . .

5. Align these lines vertically at the left.

   Quality control is important

      in reducing costs. We should

   find out where the problem is.

REVIEW EXERCISE 2-A

Part A

In the blank provided, write the word from the list that best completes each statement.

alignment	editing	paragraph
block	global	query
clarification	manipulation	symbol
dispatch	page	turnaround

1. Use a revision ___symbol___ to mark needed changes.

2. Revising a communication to improve it is ___editing___.

3. An identified segment (words, paragraphs, or pages) in a document is a/an ___blocks___.

4. Marking content for verification is called making a/an ___query___.

5. Identical changes are called ___global___ changes.

6. Changing or rearranging the message without rekeyboarding the whole message is called ___manipulation___

NAME ━━━━━━━━━━━━━━━━━━━━━━━━━━━━━ DATE ━━━━━━━━

7. Marginal notes are used for revision ___symbol___.

8. This mark (¶) indicates a new ___paragraph___.

9. The processing time to complete and return a document to the originator is called the ___Turnaround___ time.

10. Correcting uneven margins is a/an ___alignment___ revision.

Part B

In the space below, write the final copy for each edited item.

1. $122.50
      22.50
    $145.00

    $ 122.50
      22.50
    $ 145.00

2. I recommend Dr. Sharon Moore's book, <u>Time Bandit</u>.

    I recommend Dr. Sharon Moore's book, Time Bandit

3. Your game will improve with Franco Tennis Racquets.

   Your game will improve with Franco Tennis Racquets.

4. Monthly Sales Report

      July sales hit a record high. *no ¶*

      Net sales are $20,000 higher than

      July sales last year.

   Monthly Sales Report

   July sales hit a record high.

   Net sales are $ 20,000 higher than July sales last year

5. Buy a new sofa and new chair for only $298.95 $198.95 with no payments until January. Relax and enjoy your new sofa and new chair! Please hurry! Quantities are limited at this low price of only $298.95.

    Buy a new sofa for only $ 198.95 with no payments until January. Relax and enjoy your new sofa and new chair.

    Please hurry! Quantities are limited at this low price of only $ 198.95.

NAME _____ DATE _____

Part C

Mark on this postal card announcement the changes listed below. You will have blocks within blocks. To avoid confusion, label both ends of each block.

Example: $\lceil^{B}$8:15 to 10:30$\rceil^{B}$

1. Identify the announcement as block A. Then center the announcement vertically and horizontally.

2. Boldface HOUSE OF INTERIORS.

3. Underscore *At Their New Location.*

4. Block and query *8:15 to 10:30.* (Is it a.m. or p.m.?) Identify the block as B.

5. Block and identify the address as C. Then move the address to above line 5.

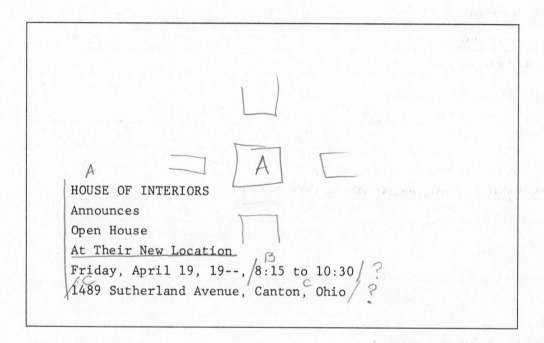

HOUSE OF INTERIORS
Announces
Open House
At Their New Location
Friday, April 19, 19--, 8:15 to 10:30
1489 Sutherland Avenue, Canton, Ohio

REVIEW EXERCISE 2-B

Two rough drafts of the same press release—one handwritten and one printed with word processing equipment—appear on pages 21 and 22. Use appropriate revision symbols to mark the same listed changes in both drafts. Symbols for global revisions will differ: Mark each occurrence on the handwritten draft. Mark only the first occurrence on the printed draft.

1. Use all-capital letters for the title.

2. Delete all three occurrences of *of Carson's.*

3. Change all three occurrences of *Brennen* to *Brennan.*

4. Insert *Thumb* before both occurrences of *Room.*

5. Insert the missing word *is* between *children* and *almost* in the third paragraph.

NAME ▬▬▬▬▬ DATE ▬▬▬▬▬

Carson's Florist, Inc.

3218 Spring Valley Drive
Spartanburg, South Carolina 29607

P
R RELEASE
E
S
S

Release: On Receipt
From: Angela Smith, President
Date: February 1, 19--

~~Carson's Florist Opens Mall Branch~~
CARSON'S FLORIST OPENS MALL BRANCH

Spartanburg, S. C., February 1-- Carson's Florist, Inc.,
opens its new branch Saturday, February 10, in the Spartan
Plaza Mall. Mr. Clayton Brennen, a recent graduate of the
University of South Carolina, is the manager of Carson's.

The mall branch of Carson's offers a unique
opportunity. In addition to buying flowers the usual way,
customers of Carson's may create their own arrangements in
the Green thumb Room. Mr. Brennen and his staff will help
when asked.

Mr. Brennen will use the Green thumb Room for flower
arranging classes this spring. His summer class
for children is almost full.

Store hours are 9 a.m. to 9 p.m., Monday through
Saturday.

Carson's Florist, Inc.

3218 Spring Valley Drive
Spartanburg, South Carolina 29607

P
RELEASE
E
S
S

Release: On Receipt
From: Angela Smith, President
Date: February 1, 19--

Carson's Florist Opens Mall Branch

Spartanburg, S.C., February 1--Carson's Florist, Inc., opens
its new branch Saturday, February 10, in the Spartan Plaza Mall.
Mr. Clayton Brennen, a recent graduate of the University of South
Carolina, is the manager of Carson's.

The mall branch of Carson's offers a unique opportunity. In
addition to buying flowers the usual way, customers of Carson's may
create their own arrangements in the Green Room. Mr. Brennen and
his staff will help when asked.

Mr. Brennen will use the Green Room for flower arranging classes
this spring. His summer class for children almost full.

Store hours are 9 a.m. to 9 p.m., Monday through Saturday.

CHAPTER

Accuracy

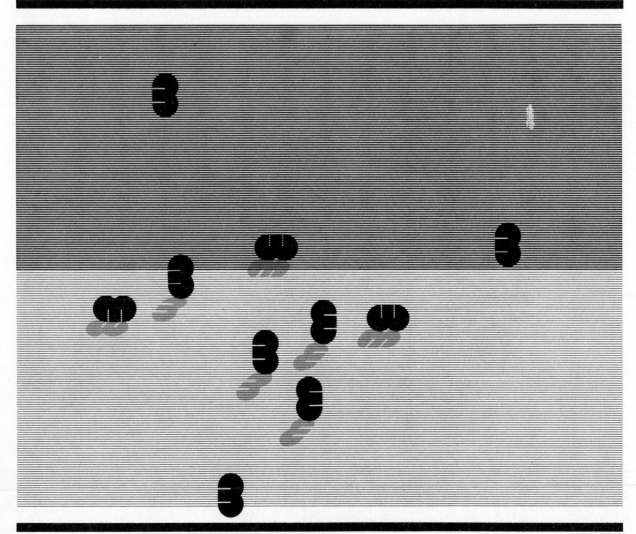

Mechanical Accuracy Is Essential

Getting the desired message across to the reader is a goal of business communications. Poor use of mechanics can get in the way of the intended message.

Have you heard a beautiful song ruined because the sound equipment wasn't working properly? You couldn't listen to the song because mechanical problems got in the way. The "poor sound" message took priority over the "beautiful song" message.

A similar thing can happen with business communications. A "poor typing" or a "poor spelling" message can take priority over an "excellent business" message.

Mechanical accuracy includes these topics:

Correct capitalization usage
Correct number usage
Typing accuracy
Correct format
Correct grammar
Correct spelling and word usage
Correct punctuation

Chapter 3 deals with editing for capitalization usage, number usage, and typing accuracy. The other topics will be covered in later chapters.

Capitalization Usage

Exercise 3-1 reviews capitalization usage. Check the Reference Section if you need a review of capitalization rules.

EXERCISE 3-1

Use revision symbols to edit for capitalization usage.

1. all orders must be placed early.

2. Each Person gave suggestions.

3. Please have the report ready by next Monday.

4. Jason lived in missouri for several years.

5. No one plans to work Christmas Eve. It is a paid holiday.

6. Your client, mr. Fred Houghton, should sign the contract.

7. Our programmers know BASIC and Fortran.

8. The best book on microcomputers is Microphobia by Dan Croy.

9. She is from the south.

10. The restaurant closes at 10 P.M.

11. Dr. Steven Candelaria, a Professor at Tryon College, invented the product.

12. His exact words were, "when can we get started?"

13. Locating the plant North of the city is a good idea.

14. Mr. Feldman works for Nelson chemical Company.

15. Sheela Mukerjee, our manager, will be here.

16. Our new sales program begins in june.

17. Please send the checks to the Bank.

18. Our company closes for the fourth of July.

19. Will you support my campaign, governor Jablonsky?

20. Dr. Sandra Odesta, our Consultant, recommended you.

EXERCISE 3-2

Use appropriate revision symbols to edit this memorandum draft for capitalization usage. When you are unsure, check the Reference Section.

MEMORANDUM

To: Ann Skouras

From: Ray Martinez

Date: November 9, 198-

Subject: Rotary Club Speaker

Thank you for recommending your former Professor, Dr. Kyle Anderson, as a speaker for rotary Club. His program will be on motivation.

The Rotary Club is eager to hear Dr. Anderson and has requested a copy of his book, Motivating Your Boss, for each member. We have permission to copy his article, "Trends in motivation," for our members.

Will you be my guest at the 12:30 P.M. luncheon on Friday, December 7, at Bishop's Cafeteria?

sk

Number Usage

Many business communications include numbers. When you are editing, check for correct number expression and verify the math.

Number Expression

Figures (*1, 2, 3*)
Words (*one, two, three*)
Roman numerals (*I, II, III*)

Math Verification

Hotel	$250.50
Plane	200.50
Total	$451.00

Exercise 3-3 reviews correct number usage.

EXERCISE 3-3

Use revision symbols to edit these paragraphs for correct use of numbers. Some paragraphs may be correct. Check the Reference Section when you are unsure of correct usage.

1. Mary Cannon, our 1st management trainee, will begin work in 2 months. She will supervise 2 high school students and 3 college students during the summer months.

2. Our sandwich shop will open the first of May with this special: two hot dogs for $.79. Shop hours will be 7:00 a.m. to 9:00 p.m., Monday through Saturday.

3. Our company style manual specifies roman numerals for chapters in a report. Tom wrote Chapter I and Chapter II. Edna wrote Chapter III.

4. Tim sold 15 vacuum cleaners and two carpet cleaners in 1 month. This qualifies him for a bonus.

5. The sweater costs $20.00; the shirt costs $10.00; and the tie costs $5.50.

    $20.00
     10.00
      5.50
    $45.50   Total

6. We spent 1,000s of dollars repairing the damage. The insurance will cover eighty percent.

7. J. B. Randolph, 23, is the youngest person ever to reach our million dollar sales plateau. He will receive his award May 25.

8. One pipe should be $2\frac{1}{4}$ feet long. The other one should be $2\frac{3}{8}$ feet long.

NAME ■■■■■■■■■■■■■■■■■■■■■■■■■■   DATE ■■■■■■■■■

9.   We need a ⅔s majority to adopt the policy. The fifty employees voted
this way: 25 for the policy, 15 against the policy, and 10 abstentions.

10.   Since January 22, 1983, we have had a 1:5 ratio of managers to supervisors.
We have reduced turnover 6 percent.

EXERCISE 3-4

Use revision symbols to edit this memo draft for correct number use. Check the
Reference Section when you are unsure of usage.

```
                            MEMO


        To:       Amos Benfield, Plant Security Officer
        From:     Karen Chan, Personnel Manager
        Date:     March 14, 198-
        Subject:  Plant Visitors


        On Friday, March 18th, at 1:00 p.m., the 2nd grade
        class from Martin Luther King Elementary School
        will visit our computer room.

        Please have 30 security clearance badges for 1
        teacher, 2 parents, and 27 children.  Maureen O'Connor,
        our safety engineer, will help you.  Do you think
        that 2 staff members can manage a group this large?

        Enclosed are 30 canteen coupons for $.49 each.  Let
        the visitors choose between soft drinks and ice cream.

        You always do a good job with school groups.  Thanks.

                            KC

        jd
        Enclosure
```

Typographical Errors

Typing accuracy is essential. Typographical errors include omissions, repetitions, incorrect spacing, floating capitals, and incorrect strokes. Study the examples below.

Omissions

> *bough the store* for *bought the store*
> *in office* for *in the office*

Repetitions

> *early inn May* for *early in May*
> *gave it it to* for *gave it to*

Omitting or repeating entire lines, pages, and paragraphs can happen very easily when you are using automated equipment to move blocks of information. Always make sure that the intended change was made.

Incorrect Spacing

> *May 5,1988* for *May 5, 1988*
> *by the   way* for *by the way*
> *onthe desk* for *on the desk*

Floating Capitals

> ^S*he is* for *She is*

Incorrect Strokes

> *form* for *from*
> *thar* for *that*
> 46 for 47

Inaccurate typing affects content. To illustrate the need for typing accuracy, think through this situation.

Suppose you are taking a course to improve office skills. What happens if you make a small, apparently insignificant error on a test? The error is typing *2:51 p.m.* instead of *2:15 p.m.* Probably, your grade will be lowered about five points. After all, a transposition is a very common error. Your test score of 95 percent sounds great.

Suppose you make the same small, apparently insignificant error on the job. In typing an itinerary for your employer, you list the flight departure time as *2:51 p.m.* instead of *2:15* p.m., the correct departure time.

What happens now? If you catch the inaccuracy and correct it before giving the itinerary to your employer, the employer has an uneventful departure. If you don't find and correct the inaccuracy, your employer may arrive at the airport just in time to catch a glimpse of the plane as it pulls away from the gate.

Typing this one figure incorrectly is the only error you made on the itinerary. You would probably get a grade of 95 percent in a classroom situation. What grade do you think an employer would give you?

Technology has made correcting errors easier. Erasers, correction paper, and liquid cover-ups are still used, but look at some automated methods.

Correcting typewriters have tape that lifts the error off the paper. The corrected text is on the final copy.

NAME ▬▬▬▬▬▬▬▬▬▬▬▬▬▬▬▬▬▬▬▬▬▬▬▬▬▬▬▬▬ DATE ▬▬▬▬▬

Electronic and memory typewriters have storage. Corrections are made on the stored item before the printing of the final copy.

Word processors and microcomputers with word processing capabilities have storage. Errors are detected on the screen, and corrections are made before the printing of the final copy.

But typographical errors must be found before any correction method may be used. Exercises 3-5 and 3-6 will give you some practice in detecting typographical errors.

EXERCISE 3-5

Compare the handwritten item with the typed item below it. Using the appropriate revision symbol, mark the errors in the typed item. If the items are the same, circle the item number.

1. *about*
   atout

2. *3:51 a.m.*
   3:15 a.m.

3. *from*
   form

4. *tomorrow*
   tommorrow

5. *accumulate*
   accummulate

6. *Mrs. Patsy Davis*
   Mrs. Patsy Davies

7. *Dr. Edward Williamson*
   Dr. Edward Williamson

8. *October 3, 1962*
   October 3, 1962

NAME ━━━━━━━━━━━━━━━━━━━━━━━━━━━━━━━━ DATE ━━━━━━━━━━

9. *June 28, 1988*
June 27, 1988

10. *243-22-8823*
243-22-8823

11. *(212) 555-8632*
(212) 555-8623

12. *$3,216.48*
$3,216,48

13. *$112,617.78*
112,617.78

14. *meeting at 3:15 p.m.*
meeting at 3:15 pm.

15. *memo from John dated*
memo form John dated

16. *accommodations are*
accommodations are

17. *The mail should be on her desk by 8:30 a.m.*
The mail should beon her desk by 8:30 a.m.

18. *Peter Zwienkelmeir wants you to call him by 8:30 p.m.*
Peter Zweinkelmeir wants you to call him by 8:30 p.m.

19. *Invoices are paid the tenth of the month.*
Invoices are paid the tenth of the month.

20. *Many fringe benefits are provided.*
Our health insurance plan is comprehensive.
Many fringe benefits are provided. Our health insurance plan is comprehensive.

NAME ━━━━━━━━━━━━━━━━━━━━━━━━━━━━ DATE ━━━━━━

EXERCISE 3-6

Too many or two few digits indicate a typographical error. Does the item have the right number of digits? Answer Yes or No. For each No, block and query the questionable part. Select the number from the list of figures below that correctly completes the item. Write the entire corrected figure. The first two items are completed as examples.

a. 3/4	h. 8223
b. .62	i. 28017
c. 10,9277	j. 45451
d. 1986	k. 1,927
e. :35	l. 82235
f. 280117	m. /79
g. 0738	n. 1644

Right No. of Digits?	Item	Corrected Figure
No	1. Phone number: (704) 555-8221 [A]?	(704) 555-8223 H
Yes	2. Social Security Number: 218-61-2392	—
YES	3. Phone number: (803) 555-8192	—
No	4. Social Security Number: 234-23-164 [A]?	234-23-1644 N
	5. Fraction: 12 2/3	
	6. Amount: 32,618 yards	
	7. Date: 12/18/7	12/18/79
	8. Money: $32.60	
	9. ZIP Code: 454541	45451
	10. Date: 12/12/85	
	11. Social Security Number: 246-81-1472	
	12. Amount: 1,9277 cases	1,927
	13. ZIP + 4: 37996-07738	
	14. Money: $85.61	
	15. ZIP + 4: 37999-0627	
	16. Time: 12:3 p.m. [A]?	12:35 P.M.
	17. Date: 11/23/87	
	18. Fraction: 2 3/ inches	
	19. Phone number: (615) 555-6212	
	20. Date: April 1, 186 [A]?	1986
	21. Time: 11:52	
	22. ZIP Code: 280177	

REVIEW EXERCISE 3-A

Part A

Using revision symbols, edit the handwritten draft of the Petty Cash Summary. Assume these facts:

1. The petty cash box contains $15.50.

2. The vouchers are correct.

No. 1201   Amount $4.50

PETTY CASH VOUCHER

Paid to *Bronson's Cafeteria*
For *Lunch with Client*

CHARGE TO

Account *Entertainment Expense*   No. 208

Received by *Lynn Pastella*   Date 5/2/8-

No. 1202   Amount $10

PETTY CASH VOUCHER

Paid to *City Office Supply Company*
For *Disk Mailing Envelopes*

CHARGE TO

Account *Office Supplies Expense*   No. 211

Received by *Ralph Starnes*   Date 5/6/8-

No. 1203   Amount $20

PETTY CASH VOUCHER

Paid to *U. S. Post Office*
For *Mailing Packages*

CHARGE TO

Account *Postage Expense*   No. 218

Received by *Justin Friend*   Date 5/7/8-

Petty Cash Summary

Beginning Balance, May 1, 198-		50	00
Entertainment Expense	20	00		
Office Supplies Expense	20	00		
Postage Expense	4	50		
Total Expenses		44	50
Balance on Hand, May 8, 198-		5	50

NAME ▬▬▬▬▬▬▬▬▬▬▬▬▬▬▬▬ **DATE** ▬▬▬▬▬▬

Part B

Use revision symbols to edit this memo draft for typographical errors and correct capitalization and number usage.

MEMORANDUM

To: Frank greene, Supervisor, Production
           Department
From: Fran Ramos, Manager
Date: April 19, 198-
Subject: Monthly Cost Report

A copy of our monthly cost report enclosed. As
you can see, your department has operated below
anticipated costs for 3 of of the 4 weeks reported.
You have met your budget and improved the quality
of your product.

Please prepare a short report relating your cost
and quality control procedures. You will have about
ten minutes ot present the report in our staff meeting Friday,April 23, at 9:30 A.M. in the conference
room.

                    FR

jb
Enclosure

REVIEW EXERCISE 3-B

Using revision symbols, mark all errors in each purchase requisition. (See pages 34 to 36.) If the requisition is correct, put a check next to the requisition number. The memo-reply forms, the date needed, and the date requested are correct.

NAME ━━━━━━━━━━━━━━━━━━━━━━━━━━━━━━━━ **DATE** ━━━━━━━

MEMO

To *Ilene Marks*   Subject *Desk and Chair*

Did you choose a desk and chair for your office?

Please list the catalog number and the supplier's name and address.

From *John Whitesides*   Date *4/15/8—*

- -

REPLY

Please order: Desk, Catalog no. 3286 (Walnut)

Chair, catalog no. 3186 (Green)

Supplier: Office Interiors, Inc.

Post Office Box 1508

Shreveport, LA 71165

From *Ilene Marks*   Date *4/16/8—*

PURCHASE REQUISITION

To Sally Fisher, Purchasing Agent   Requisition No. P-2107
From John Whitesides   Date Requested 4/16/8—
Department Personel   Date Needed 4/16/8—
Deliver To Ilene Marks, Building E, Office 201   Account No. 221

Quantity	Description	Suggested Source
1	Desk, Catalog No. 3268 (Walnut)	Office interiors, Inc. Post Office Box 1508 Shreveport, LA 711655
1	Chair, Catalog No. 3186 (Green)	

Reason for Request   New employee

Authorized Signature *John Whitesides*

Below Line—For Purchasing Department Use Only

Ordered From

Purchase Order No. _____

Date Ordered _____

Date Received _____

Purchasing Agent

NAME ▬▬▬▬▬▬▬▬▬▬▬▬   DATE ▬▬▬▬▬▬▬

MEMO

To _Andy Evans_   Subject _Word Processing System_

Management approved the word processing system.
Please list the description and the supplier's
name and address.

From _John Whitesides_   Date _4/16/8-_

REPLY

Approval for the word processing system is good news!
Description: Quick Word Processing System, Model QK-2
Supplier: Quick Systems, Inc.
Post Office Box 14502
Baton Rouge, LA 70808

From _Andy Evans_   Date _4/17/8-_

PURCHASE REQUISITION

To Sally Fisher, Purchasing Agent   Requisition No. P-2108
From John Whitesides   Date Requested 4/18/8-
Department Personnel   Date Needed 4/18-8-
Deliver To Andy Evans, Building E, Office 218   Account No. 221

Quantity	Description	Suggested Source
1	Quick Word Processing System Model QK-2	Quick Systems, Inc. Post Office Box 14502 Naton Rouge, LA 70808

Reason for Request ___Increase paperwork___

Authorized Signature _John Whitesides_

Below Line—For Purchasing Department Use Only

Ordered From   Purchase Order No. _____

Date Ordered _____

Date Received _____

Purchasing Agent

MEMO

To *John Whitesides*   Subject *Application Forms*

Please requisition 500 application forms, Reorder No.
R-20631A, for delivery by May 1. Order from this
company: Office Printing and Supply
1803 Crawford Road
Shreveport, LA 71107

From *Ann Barnes*   Date *4/18/8-*

- -

REPLY

The requisition was sent to purchasing today.

From *John Whitesides*   Date *4/18/8-*

PURCHASE REQUISITION

To Sally Fisher, Purchasing Agent   Requisition No. P-2109
From John Whitesides   Date Requested 4/18/8-
Department Personnel   Date Needed 5/1/8-
Deliver To Ann Barnes, Building E, Office 212   Account No. 212

Quantity	Description	Suggested Source
5000	Application Forms, Reorder No. R-2063A	Office Printing and Supply 1803 Crawford Road Shreveport, LA 71107

Reason for Request ____Supply islow._____

Authorized Signature *John Whitesides*_____

Below Line—For Purchasing Department Use Only

Ordered From   Purchase Order No. _____

_____   Date Ordered _____

_____   Date Received _____

_____

                                        Purchasing Agent

NAME ▬▬▬▬▬▬▬▬▬▬▬▬▬▬▬▬   DATE ▬▬▬▬▬

CHAPTER

Details

Details Are Necessary

Have you ever received a message that left out a very important detail? For example, maybe you were invited to a party. All details were given except the time.

An incomplete message often occurs because the writer is so familiar with the message that omitted details are not obvious. Omitted details are, however, obvious to the reader.

The next two exercises will give you practice in editing for completeness in business communications.

EXERCISE 4-1

Indicate the detail from the list below that would appropriately complete each numbered item by placing the letter of the detail in the blank provided. Use each letter only once.

a. at 3:30 p.m.
b. p.m.
c. in our recreation room
d. size 16
e. (201)

f. at the Lunch Box Cafeteria
g. by Speedy Package Delivery Service
h. for president
i. note pads
j. Illinois 60453

_____C_____ 1. Our fitness class will meet each morning (Monday through Friday) from 6:30 to 7:30.

_____F_____ 2. Let's have lunch together on Monday, April 5. I'll meet you at noon.

_____H_____ 3. The nominating committee for the Executive Dinner Club completed its recommendations for next year's officers. Will you accept the nomination?

_____A_____ 4. All office personnel will meet in our conference room on Wednesday, October 4.

_____i_____ 5. Order these three items: pens and pencils.

_____B_____ 6. Dr. Wolinsky's presentation to all shift supervisors will be at 8:30 in the auditorium.

_____J_____ 7. Please ship this order by UPS to Manfred Manufacturing, Inc., 1127 Alexander Plaza, Oak Lawn.

_____E_____ 8. Our California phone number is (916) 555-8327, and our New Jersey number is 555-9726.

_____G_____ 9. Your order, No. 2323, was shipped on September 23 to your home address. You should have received it six weeks ago.

_____D_____ 10. I would like to order two white shirts, Catalog No. 2876-98. Please ship them by UPS to J. C. Paul, 123 Red Coach Road, Dayton, OH 45429.

EXERCISE 4-2

Each item asks for information. Circle the letter answer that includes all the facts needed for a reply.

1. What is our warehouse inventory on these model numbers: 2361 (radio), 2846 (stereo), 1287 (calculator), and 5682 (table)?

   a. 10 No. 2846 (stereo)
      15 No. 1287 (calculator)
      20 No. 5682 (table)
      20 No. 2361 (radio)

    b. 20 No. 2361 (radio)
       10 No. 2846 (stereo)
       15 No. 1287 (calculator)
       20 No. 2361 (radio)

    c. 20 No. 2361 (radio)
       10 No. 2846 (stereo)
       15 No. 2846 (stereo)
       20 No. 5682 (table)

2. Please list your name, address, phone number, and social security number.

    a. J. D. Hand, Jr., 1159 Tartan Lane, Pueblo, Colorado 81001
       (303) 555-6728

    b. J. D. Hand, Jr., 1159 Tartan Lane, Pueblo, CO 81001
       (303) 555-6728, 322-70-5728

    c. J. D. Hand, Jr., 1159 Tartan Lane, Pueblo, CO 81001
       322-70-5728, 322-70-5728

3. How many clerks, receptionists, secretaries, and word processing operators does your company employ?

    a. 18 clerks, 12 receptionists, 10 word processing operators

    b. 18 clerks, 5 secretaries, 11 stenographers, 10 word processing operators

    c. 10 word processing operators, 12 receptionists, 5 secretaries, 18 clerks

4. Please tell me the business hours and location of your retail store. Do you accept credit cards and out-of-state checks? Do you sell clothes for men, women, and children?

    a. Located at 134 Murphy Drive, Pawtucket, Rhode Island 02864.
       Attractive clothes for the whole family.
       Hours from 8:30 a.m. until 4:30 p.m., Monday through Friday.
       Do not accept out-of-state checks.

    b. Attractive clothes for the whole family.
       Located at 134 Murphy Drive, Pawtucket, Rhode Island 02864.
       Do accept major credit cards; do not accept out-of-state checks.
       Hours from 8:30 a.m. until 4:30 p.m., Monday through Friday.

    c. Located at 134 Murphy Drive, Pawtucket, Rhode Island 02864.
       Do accept major credit cards; do not accept out-of-state checks.
       Hours from 8:30 a.m. until 4:30 p.m., Monday through Friday.

5. Do your employees have health and accident insurance, dental insurance, and life insurance? How much paid vacation do employees get after working with you for one year? Do employees have a retirement plan? Do you offer other fringe benefits?

    a. Employees have health and accident, dental, and life insurance.
       Employees have a retirement plan.
       Employees have one week of paid vacation after working here one year.

    b. Employees have a retirement plan.
       Employees have one week of paid vacation after working here one year.
       Employees have health and accident, dental, and life insurance.
       Employees have no additional fringe benefits.

    c. Employees have health and accident insurance.
       Employees have dental insurance.
       Employees have life insurance.
       Employees have one week of paid vacation after working here one year.
       Employees have a retirement plan.

NAME ━━━━━━━━━━━━━━━━━━━━━━━━━━━━━ **DATE** ━━━━

Complete Format

Another part of editing is making certain that formats are correct and complete. Example Formats 1 to 12, for memos, letters, and reports follow. Review the formats, study the following pages, and complete the accompanying exercises.

MEMORANDUMS

Memorandums are used for interoffice communication. Many organizations provide printed forms for memorandums, but it is also acceptable to type them on plain paper. How you set the left margin on a memo will depend on the format used for the guide words. (The guide words have been set in initial capital only as this is easier for automated equipment.) Notice the left margin in the two memo formats shown here. In Format 1, the left margin is aligned with the beginning of the guide words. In Format 2, the left margin is set two spaces after the colon following the guide words.

When typing a memo on a printed form, set the right margin approximately equal to the left margin. When typing a memo on plain paper, set the left and right margins for a 5- or a 6-inch line depending on the length of the memo.

Heading centered on line 7	↓7 MEMORANDUM ↓3

Guide words	To: Liza Berger, Supervisor
	          Production Department
	          ↓2
	From: Maurice Lemaire, Manager
	          ↓2
	Date: June 20, 198-
	          ↓2
	Subject: Vacation Schedule
	          ↓3

Body	Xxx xxxx xxxx xxxxx xxxxxx xx xxxx. Xxxx xxxx xx
	xxxxxx. Xxx xxxx xxxxx xx xxxx xxx xx xx. Xxx x
	xxx xxxx xxxxx xx xxxx xxxxxx.
	          ↓2
	Xxxxx xxxx xx xxxxxx xxxxx. Xxx xxxx xxxx xxx xxxxx
	xxx xxxxxx xx xxxx. Xxx xxxx xxx xxxx xxx xxxxxx.
	          ↓2

Writer's initials begin at center	Xxxx xxxx xx xxxxx xx xxxxxx xx xxxx. Xxx xxx xxx
	xxxxxx xxxx xxx xx xxxxxx xxxx xxxx.
	          ↓2
	              ML
	              ↓2

Reference initials	kb

MEMORANDUM ON PLAIN PAPER
Format 1

Printed
heading
centered

M E M O R A N D U M
↓ 3

Printed
guide
words

To: Fran Greene, Supervisor, Shipping Department

From: Fred Richards, Manager

Date: April 19, 198-

Subject: Vacation Schedule
↓ 3

Body

Xxx xx xxxx xx xxx xxxxxx xxxx xxx. Xxxxx xxx xxx
xxxx. Xxxxx xxxx xxxxx xxxxxx xxxxxx. Xxxxx xx x
xxxxxx xxxxxx xxxx xxxx xxxxxxx.
↓ 2

Xxxxx xxxx xxx xx xxxxx xxxx xxxxxx. Xxxxx xxx xx
xxxxxx xxxxx xxxx xx xxx. Xxxx xxxxx xxxx xxx.
↓ 2

Writer's
initials begin
at center

Xx xxxxxx xxx xxxx xxxxxx xx. Xxxxxx xxxxxx xx xx
xxxxx xxx. Xxxxx xxxx xxxxx xxxxxx xxx xx.
↓ 2

FR
↓ 2

Reference
initials

kb

MEMORANDUM ON PRINTED FORM
Format 2

EXERCISE 4-3

Complete the following statements about memorandums.

1. List the guide words.

   a. _____

   b. _____

   c. _____

   d. _____

2. The body of a memorandum is _____-spaced.

3. Return the carrier _____ times after the subject line.

4. Return the carrier _____ times between paragraphs.

5. Return the carrier _____ times before and after the writer's initials.

6. On the plain paper memo, the left margin of the body is aligned at the left with the beginning of

   the _____ _____.

7. On the printed memo form, the left margin for the whole memo is set _____ spaces after

   the colon following the guide words.

NAME ━━ DATE ━━━━

LETTERS

SOCIAL-BUSINESS LETTER

The social-business letter format is used by business people when company letterhead is not appropriate.

Format 3 shows a social-business letter in modified-block style (date and closing lines at center) with standard punctuation (a colon after the salutation and a comma after the complimentary closing). However, block style with open punctuation (see Format 6) may also be used.

SIMPLIFIED LETTER

The simplified business letter format (Format 4) was designed for efficiency. All lines are flush with the left margin, and the salutation and complimentary closing are omitted.

Notice that the subject line (triple-space before and after) and the writer's name and title are in all capitals.

Writer's address begins at center on line 13

Date

    3829 Oxford Road
    Springfield, OH 44904
    March 10, 198-
                    ↓5

Inside address

    Ms. Phyllis Jamison
    8827 Lakecrest Drive
    Springfield, OH 44903
                    ↓2

Salutation

    Dear Ms. Jamison:
                    ↓2

Body

    Xxxx xxxxx xx xxxx xxxxx xx xx. Xxxx xxxx xx
    xx xxxxx xxx. Xxx xxxx xxx xxxx xxx xxxxx xxxxx x
    xxx. Xxx xxxx xx xxxx xxxxx xxxxxx.
                                    ↓2

    X xxxx xxx xx xxxx xxxxxx xxxx xxxx. Xxxx xx
    xxxx xxxxxx xxxxx xxxx xxx xxxxx xxxx. Xxx xxxxxx
    xxx xxxx xxxxxx xxxx xxxx xx xxx. Xxx xxxx xxx xx
    xxx. Xxx xxxx xx xxxxxx xx xxxx xxxxxx xx. Xxxxx
    xxxxx xxxxxx xxxxx xxx xxxxx xxxxx.
                                    ↓2

    Xxxxx xxxx xxx xxxxxx xxx xxxx xxxxxx xxx xxx
    xxxx xxxxxx xxx xxxx. Xxxxx xxx xxxx xxxxx xxxxxx
    xxx. Xxxx xxxx xx xxxx xxx xxxx xxx xxxx xx xxxxx
    xxxxxxxxxx. Xxxxx xxxxx xxx xxxx xxx.
                                    ↓2

                    Sincerely yours,
                                    ↓4

Complimentary closing begins at center

                    Dana Ramsey
                    Dana Ramsey

SOCIAL-BUSINESS LETTER IN MODIFIED—BLOCK STYLE
STANDARD PUNCTUATION
INDENTED PARAGRAPHS
Format 3

NAME ▬▬▬▬▬▬▬▬▬▬▬▬▬▬▬▬ DATE ▬▬▬▬

Data Company, Inc.

3281 Kenan Drive • Flint, Michigan 48504 • (313) 555-6391

Date on
line 15   May 10, 198-
                                   ↓5

Inside   Mr. Ronald Lee
address   690 Bellevue Avenue
                       Flint, MI 48506
                                   ↓3

Subject
line   PROGRAMMING POSITION OPEN
                                   ↓3

Body   Xxx xxxx xxx xxxxxx xx xxxx xx xxxx xxxxxx xxx xx.
                       Xxx xx xxxxxx xxxx xxxxxxx xxxx xxxxxxxxxx.
                                                             ↓2
                       Xxxxx xxxx xxxxx xxx xxxxx xx xxxxxx xxxx. Xxxxxx
                       xxx xxx.   Xxxxxx xxxx xxxxx xxx xxxx xxx xxxx xxxx.
                       Xxxx xxx xxxxx xxxxx xxxxxx.
                                                   ↓2
                       Xx xxxxxx xxxx xxxxx xxxxxxxx xxx xxxxx xx xxxxx.
                       Xxxxx xxxxx xxxxx xx xxxx xxxxx xxxx xxxx.
                                                             ↓5

                       Frank Luciano

Writer's
identification   FRANK LUCIANO - PERSONNEL MANAGER
                                                     ↓2
Reference
initials   dp

SIMPLIFIED LETTER
OPEN PUNCTUATION
Format 4

EXERCISE 4-4

Write the correct answers in the blanks provided. If you are in doubt, refer to the style letters or the text explanation.

Social-Business Letter

1. The social-business letter doesn't have a letterhead; therefore, the writer's _____ must be typed above the date.

2. Writers often type their own social-business letters; therefore, no typist's _____ are shown.

3. In standard punctuation, there is a _____ after the salutation.

4. A _____ follows the complimentary closing in standard punctuation.

5. In modified-block style, the writer's address, the date, and the closing lines begin at the _____.

Simplified Letter

6. All lines are typed _____ with the left margin.

7. The simplified format omits two usual letter parts: the first is the _____.

8. The second letter part that is omitted is the _____.

9. Two lines are typed in all capitals: the first is the _____.

10. The second line typed in all capitals is the _____ line.

TWO-PAGE LETTERS: BLOCK AND MODIFIED-BLOCK STYLES

In block letter style, all lines begin at the left margin. In modified-block letters, the date line and the closing lines begin at center; all other lines begin at the left margin. Paragraphs may be indented or not indented in modified-block style.

A two-page format is like a one-page format except that the second page is identified by a heading and does not contain a letterhead, inside address, or salutation.

Block-Style Heading

Reynolds Company
Page 2
October 21, 198–

Modified-Block Style Heading

Reynolds Company   2   October 21, 198–

The second-page heading must agree with the inside address. Look at the modified-block letter (Format 5). The inside address shows that the letter is written to Mr. Joe Bianco. Both the salutation (*Dear Mr. Bianco*) and the second-page heading (*Mr. Joe Bianco*) agree with the inside address.

Now look at the block-style letter (Format 6). Even though this letter was written to the attention of Mr. Joe Bianco, the letter was addressed to Reynolds Company. *Reynolds Company* is used in the inside address and in the second-page heading. Using Mr. Bianco's name in the inside address, salutation, and second-page heading would be incorrect. (Because the letter is addressed to a company, the salutation should be plural; *Ladies and Gentlemen* would be appropriate.)

When a paragraph is continued from the first to the second page, at least two lines must be left on the first page. At least one full and one partial line must be continued onto the second page. Words should never be divided between pages.

4736 Newport Drive
Flint, Michigan 48504
(313) 555-8745

Date on
line 15
                                      October 21, 198-
                                                   ↓5

Inside   Mr. Joe Bianco
address   624 Bellevue Avenue
              Brockton, MA 02402
                                 ↓2
Salutation   Dear Mr. Bianco:
                               ↓2
Body   Xxxxx xxx xxxxx xx xxxxx xxxxx xxxx. Xxxx xx
              xxxxxxxx xxxx. Xxx xxxxxx xxxx xxxxxx xxxxx.
                                                        ↓2
Divided   Xxxxx xxxxx xxxxxx xxxx xxxx xxxxxx xxxxxx xx
paragraph: xxxx xxxxx xxxxx. Xxxxx xxxxx xxxx xxxxx xxx.
at least 2
lines on
first page   ∼∼∼∼∼∼∼∼∼∼∼∼∼∼∼∼∼∼∼∼∼∼∼∼∼∼∼∼∼∼∼∼∼

                  Xxxxx xxx xxxxx xx xxxxxx xxxx xxxx xxxxxxxxx
              xxxx xxxxxxxx. Xxxxx xxxx xxxx xxxx xxxx xx . . .

Second-page
heading on
line 7   ┌──┐
              │ Mr. Joe Bianco   2   October 21, 198-   │
                                                              ↓3
Continued   xxxx xxxxx xxxxx xx xxxxx. Xxxxx xxxx xxx xxxxxxx
paragraph: xxxx xxxx xxxxxxxxxx.
at least a full   ↓2
and a partial line   Xxxxx xxxx xxxxx xx xxxxxx xxxx xxxx xxxxxxxx
on second page xxx. Xxx xxxx xxxx xxxxxx xxxxx xxx xxxxxx.
                                                         ↓2
Complimentary   Sincerely yours,
closing   ↓4
                              Anna J. Goldberg

                              Anna J. Goldberg
                              Manager, Building Operations
                                                          ↓2
Reference
initials   mt

**TWO-PAGE BUSINESS LETTER, MODIFIED-BLOCK STYLE
STANDARD PUNCTUATION
INDENTED PARAGRAPHS
Format 5**

NAME ▬▬▬▬▬▬▬▬▬▬▬▬▬▬▬ DATE ▬▬▬▬▬

4736 Newport Drive
Flint, Michigan 48504
(313) 555-8745

Date on line 15	October 21, 198- ↓5
Inside address	Reynolds Construction 624 Bellevue Avenue Brockton, MA 02402 ↓2
Attention line	Attention: Mr. Joe Bianco ↓2
Salutation	Ladies and Gentlemen ↓2
Message	Xx xxxxx xxxxxx xxxxxx xxxx xxxxxx xxxx xxxx xxxxx xxxx. Xxxx xxxx xx xxxxx xxx xxxx xxxx.
Divided paragraph: at least 2 lines on first page	Xxxxxx xxxx xxxxxx xxxx xxxx xx xxxxx xxxx xxxx xx xxx. Xxx xxxx xxxx xxx xxxxx xxxx xxx xxxx xxxxx.
	Xxx xxxxxx xxxx xxxxx xxxxxx xxxxxxxxx xxx xxxxxxx xxxxxxx xxxx xxxx. Xxxx xxxxx xxxxxx xxxxxx . . .
Second-page heading on line 7	Reynolds Construction Page 2 October 21, 198- ↓3
Continued paragraph: at least a full and partial line on second page	xxxxx xxxxx xxxx. Xxxxx xxxxx xxxxxxxxx xxxxx xxx xxx xxxx xxxxx xxxx.
	Xxxxx xxxx xxxxx xxx xxxx xxxx xxxxx xx xxxxxxxxxx xxx xxxxxx xxxx xxx x xxxxx xxxx xxx xxxxx. ↓2
Complimentary closing	Sincerely yours *Anna J. Goldberg* ↓4
	Anna J. Goldberg Manager, Building Operations ↓2
Reference initials	mt

TWO-PAGE BUSINESS LETTER, BLOCK STYLE
ATTENTION LINE
OPEN PUNCTUATION
Format 6

NAME ▬▬▬▬▬▬▬▬▬▬▬▬▬▬▬   DATE ▬▬▬▬

EXERCISE 4-5

Complete the following items about two-page letters. If you are in doubt, refer to the style letters or the text explanation.

1. Which two letter parts must agree with the inside address?

   a. _____

   b. _____

2. Show the salutation and the possible second-page headings for a letter with the following inside address. The letter was dated May 1, 198–.

   > Ms. Adele Rheinhold
   > Apex Marketing Company
   > Post Office Box 2821
   > Ogden, UT 84404

   a. Salutation (block style)

   _____

   b. Second-page heading (block style)

   _____

   _____

   _____

   c. Second-page heading (modified-block style)

   _____

   _____

   _____

3. Show the salutation and the possible second-page headings for a letter with the following inside address. The letter was dated May 1, 198–.

   > Apex Marketing Company
   > Post Office Box 2821
   > Ogden, UT 84404
   > Attention: Ms. Adele Rheinhold

   a. Salutation (modified-block style)

   _____

   b. Second-page heading (block style)

   _____

   _____

   _____

c. Second-page heading (modified-block style)

_____

_____

_____

_____

4. At least _____ lines of a divided paragraph must be left on the first page.

5. At least _____ lines of a divided paragraph must be continued on the second page.

LETTERS WITH SPECIAL NOTATIONS

Listed here are the most frequently used notations. Notice the placement of the notations in Formats 7 and 8.

Mailing notations record any special mailing such as certified mail, special delivery, and express mail.

Attention lines direct letters addressed to a company to specific persons within that company.

Subject lines briefly identify the main topic.

Enclosure notations show that something is included with the letter. Use *Enclosure* if one item is enclosed; use *Enclosures* if more than one item is enclosed. The specific enclosure or the number of enclosures can be shown.

Enclosure: Lease Agreement
Enclosure
2 Enclosures
3 Enclosures
1 Enclosure

Copy notations show the distribution of copies. The initials *cc* (for *carbon copy*) are usually used to introduce this notation even when not carbons but other kinds of copies (such as electronically printed or photostatic copies) are used. Some writers prefer to use a single *c*. Others now use the *cc* to mean *courtesy copy* instead of *carbon copy*.

Blind copy notations are used when the writer sends someone a copy of the letter without the addressee's knowledge. The notation goes on the file copy— *not* on the original. It may go on other copies as specified by the writer. For example, the writer may want the blind copy notation to appear on Mr. Cordova's copy but not on Mrs. Soo's copy.

Postscripts are comments added after the letter has been typed. They are indented or not indented depending on the letter style used.

NAME ▬▬▬▬▬▬▬▬▬▬▬▬▬▬▬▬▬▬ DATE ▬▬▬▬

MANAGEMENT CONSULTANTS, INC.

3699 FAIRWOOD ROAD • CINCINNATI, OHIO 45239 • (513) 555-6823

Date on line 15	June 2, 198- ↓5

Inside address
```
Executive Office Services
5601 Dulap Road
Cincinnati, OH 45247  ↓2
```
Attention line
```
Attention:  Ms. Pat Demay  ↓2
```
Salutation
```
Ladies and Gentlemen:  ↓2
```
Subject line
```
Subject:  Electronic Equipment Exhibition  ↓2
```
Body
```
Xxxx xxxxxx xxxx xxxxxx xx xxxxxxx xxxx xxx xxxxxx
xxxxxx.  Xxxxx xxxxx xxxx xxx xxxx xxx.

Xx xxxxx xxxx xxxxxx xxxxxx xxxx xxxx xxxxx xxxxxx
xxxx.  Xxxxx xxxxx xxxx xxx xxxxx xxxxxx xxxxxx.
Xxxxxx xxx xxxxx xxxxxx x xxxxxx xxxx xxxx.

Xxxxx xxxxx xx xxxxx xxxxxx xxxx xxxxxx x xxx xxxx
xxx xxxxx xxxx.  Xxxxx xxxxx xxxxx xxxx xxx.
```
Complimentary closing
```
                    Sincerely yours,  ↓2
```
Company name
```
                    MANAGEMENT CONSULTANTS, INC.  ↓4


                    Logan Bolinsky
                    Senior Consultant
```
Reference initials
Enclosure notation
Mailing notation
```
rk
2 Enclosures
Certified
```
Copy notation
```
cc:  Mr. Esteban Cordova
     Mrs. Lillian Soo  ↓2
```
Postscript
```
PS:  Xxxxxx xxx xxxxxx xxx xxxxx.  Xxxxxx xxx xxxx
xxx xxxxxxxx xxxxx.
```
Blind copy notation on copy only

**SPECIAL NOTATIONS
BUSINESS LETTER, MODIFIED-BLOCK STYLE
STANDARD PUNCTUATION
Format 7**

"COPY"
often printed or
typed on copies

COPY

June 2, 198-
↓5

Executive Office Services
5601 Dulap Road
Cincinnati, OH 45247

Attention: Ms. Pat Demay
↓2

Ladies and Gentlemen:
↓2

Subject: Electronic Equipment Exhibition
↓2

Xxxx xxxxxx xxxx xxxxxx xx xxxxxxx xxxx xxx xxxxxx
xxxxxx. Xxxxx xxxxx xxxx xxx xxxx xxx.
↓2

Xx xxxxx xxxx xxxxxx xxxxxx xxxx xxxx xxxxx xxxxxx
xxxx. Xxxxx xxxxx xxxx xxx xxxxx xxxxxx xxxxxx.
Xxxxxx xxx xxxxx xxxxxx x xxxxxx xxxx xxxx.
↓2

Xxxxx xxxxxx xx xxxxx xxxxxxx xxxx xxxxxx x xxx xxxx
xxx xxxxx xxxx.   Xxxxx xxxxx xxxxx xxxx xxx.
↓2

                                  Sincerely yours,
↓2

                                  MANAGEMENT CONSULTANTS, INC.
↓4

                                  Logan Bolinsky
                                  Senior Consultant
↓2

rk
2 Enclosures
Certified

cc: Mr. Esteban Cordova
     Mrs. Lillian Soo
↓2

PS: Xxxxxx xxx xxxxxx xxx xxxxx. Xxxxxx xxx xxxx
xxx xxxxxxxx xxxxx.
↓2

bcc: Mr. Drew Loftis

Blind copy
notation
double
spaced below
the last item

FILE COPY
SPECIAL NOTATIONS
BUSINESS LETTER, MODIFIED-BLOCK STYLE
STANDARD PUNCTUATION
Format 8

NAME ▬▬▬▬▬▬▬▬▬▬▬▬▬▬▬▬   DATE ▬▬▬▬▬▬▬

EXERCISE 4-6

Use the following words to complete the fill-in items below.

anonymous	enclosure
attention line	mailing
blind copy	photostatic
carbon	postscript
copy	specified line
electronically printed	subject line

1. *Certified* is an example of a/an _____ notation.

2. The main topic is often expressed in a/an _____ .

3. A/An _____ notation indicates on the original who will receive copies.

4. Use a/an _____ notation to show that something is included with the letter.

5. Use a/an _____ to direct a letter to a specific person within a company when the letter is addressed to the company.

6. A/An _____ notation specifies persons receiving copies without the addressee's knowledge.

7. A comment added after the letter is typed is a/an _____ .

8. List three kinds of frequently used copies.

   a. _____

   b. _____

   c. _____

REPORTS

Definitely, reports play an important role in business operations. Supervisory people need reports to assure quality and efficiency. Management needs reports for decision making. Financial departments require reports for accounting purposes and other reasons. Lenders and stockholders want reports on their interests in the company. The list could go on.

Business reports can be formal or informal as need or company policy dictates. See Formats 9 to 12 for some basic report typing considerations.

All reports should be given a title and include a by-line (name of the person who wrote the report). Depending on the length and organization of the report, headings, sub-headings, and enumerations may be required. Additional elements, such as documentation, headers, and footers, may be necessary in a formal report.

DOCUMENTATION

Business reports may or may not have **documentation** (footnotes and a bibliography). A proposal to buy a computer may have documentation supporting the reasons for a specific recommendation. A report on sales calls made would not have formal documentation.

Documentation may be **footnotes** at the bottom of the page or **notes** at the end of the report (sometimes called **endnotes**).

A report may also have a **bibliography**—an alphabetized listing of sources.

NAME _____   DATE _____

Title centered on line 13	MICROCOMPUTERS ↓2
Subtitle or By-Line	By Freida Griffen ↓3
	Xxxxx xxxxxxx xxxx xxx xx xxxx xx xxxxx xx xxxxx xxxxx
Superscript footnote number	xxxx xxxxxxxxxxx. Xxxx xxxx xxxxx xxxxxx xxxxxxx. Xxxxxxx xxxxxxx xxxxxxx xxxxxxx xxxxxx xxxxxxxxxx xxxxxx.¹ ↓3
Side heading	EQUIPMENT ↓2
	X xxx xxxx xx xxxxxxx x xxxx xxxx xxxxxx. Xx xxx xxxx xxxx xxxx xxxxx. Xxx xxxx xxx xxxx x xxxx xx xxxxxx. ↓2
Paragraph heading	Brands of Computers. Xx xxx xxxxxx xxxxx xx xxxxxxxxx xxxx xxxxxx xxxx xxx xxxx.² Xxxxx xx xxxxx xxxxxxx xxxxxx xxx. Xx xxxx xxxxx xx xxxxxx xxxx xxxxxxxx xxxxx. ↓2
Enumerations: Indent 5 spaces from both margins	1. Brand A has xxxx xxx xxxx xxx xxxx xxxxx xxxxx xxxxxxxx. Xxx xxxx xxxxx xxxx xxxx xxxxx. ↓2
	2. Brand B has xxx xxxxxx xxxxxx xxxxx xx xx x xx xxxxxxxxxxxxx. Xxxx xxxx xxxx xxxx xxxx xxxx xxxxx.
Divided paragraph: at least 2 lines on first page	3. Brand C has xxx xxxx xxx xxx xxxxx xxxxxx xxxx xxxxx xxxxxxxxx. ↓2
	Printers. Xx xxx xxxx xxxxx xxx xxxxxxxxx xxx xxxx xx xxxxxxxx. Xxx xxxx xxxxx xxx xxxxx xxx. Xxxx xxxx xxxx x ↓1
Footnote dividing line	——————————————— ↓2
Footnotes	¹Ronald Davis Franco, Computing Made Simple, Henley Publishing Company, Inc., New York, 1983, p. 234.
	²Linda H. Solomon, "Microcomputing for Offices," Automation Trends, November 1982, pp. 43-45.
Bottom margin of 6 to 9 lines	

**REPORT: FIRST PAGE
WITH FOOTNOTES AT BOTTOM OF PAGE
Format 9**

Page number
on line 7

2

Continued
paragraph:
at least one
full and one
partial line

xxxx.   Xxxx xxx xxx xxxx xxx xxxxx xxxx.³ Xxxx xxx xxxx xxx

xxxx xxxxxx xxxxxx. ↓3

Side
heading

USES OF MICROCOMPUTERS ↓2

Xxxx xxxx xx xxxxx xx xxxxx xx xxxxx xx xxxx. Xxxx xxx

xxxx xxx xxxxx xx xxxxxx xx xxxxx xx xxxx xxx. ↓2

Paragraph
heading

<u>Accounting Records</u>. Xxxx xxxxx xx xxxxxxx xxxxxxx xxxx

xxxx xxx xxxxx. Xxxx xxxxx xxx xxxxx xxxxx.

<u>Personnel Records</u>. Xxx xxxx xx xxxxx xxx xxxxxxxx xxxx

xxxx. Xxxx xxxxx xxx xxxxxx xx xxxxxx xx xxxxx xxxx xxxx.

<u>Word Processing</u>. Xxxxxx xxxxxx xx xxxx xxx xxxx xxxx x

xxxx xxxxxxxx. Xxx xxxx xxxxx xxx xxxx xxxxx xxxxxx xxxxx.⁴ ↓2

Xxx xxxxx xx xxxxx xx xxxxxx xxx xxxx xxxxx xx xxxxxxxx

xxxxxx. Xx xxx xxxx xxxxx xxxxx xxx xxxxxx xxxx xxxxx xxxx

xxx xxxxxx xxx xxxx xxx xxxxxx xxx xxxx.⁵

Add space above
footnote dividing
line to keep 6-9
line margin

Footnote
Dividing line:
2 inches long _____

³Renaldo S. Gonzales, "Micros for Your Office," <u>Data
Magazine</u>, January 1983, pp. 67-69.

⁴Ibid., p. 70.

⁵Solomon, p. 46.

**REPORT: LAST PAGE
WITH FOOTNOTES
Format 10**

NAME ▬▬▬▬▬▬▬▬▬▬▬▬▬▬▬▬▬▬▬▬▬▬▬ DATE ▬▬▬▬▬▬

Header
on line 7   MICROCOMPUTERS by Freida Griffen   2 ↓3

Continued
paragraph: xxxx. Xxxx xxx xxx xxxx xxx xxxxx xxxxx.³ Xxxx xxx xxxx xx
at least one
full and one   xxxxx xxxxxxxx xxxxxxx. ↓3
partial line

Side   USES OF MICROCOMPUTERS
heading
                      Xxxx xxxx xx xxxxx xx xxxxx xx xxxxx xxx xxxxx. Xxxx x

               xxxx xxx xxxxx xx xxxxxx xx xxxxxx xx xxxx xxx. ↓2

Paragraph   Accounting Records. Xxxx xxxxx xx xxxxxxxx xxxx xxxxxx
heading
               xxxxx xxx xxxxx. Xxxx xxxxx xxx xxxxxx xxxxx.

                      Personnel Records. Xxx xxxx xx xxxxx xxx xxxxxxxx xxxx

               xxxx. Xxxx xxxxx xxx xxxxxx xx xxxxxx xx xxxx xxxxx xxxx.

                      Word Processing. Xxxxxx xxxxx xx xxxx xxx xxxx xxxx xx

               xxxx xxxxxx. Xxx xxxx xxxxxx xxx xxxx xxxxx xxx xxxx xxxx.⁴ ↓2

                      Xxxx xxxx xx xxxxxx xx xxxxxx xxx xxxx xxxxx xx xxxxxxx

               xxxxxxx. Xxxxx xxxx xxxxx xxxxxx xxx xxxxxx xxxx xxxxxx xxx

               xxx xxxxxx xxx xxxx xxx xxxxx xxx xxxx.⁵

**INFORMAL REPORT: LAST PAGE
WITH HEADER
ENDNOTES AND BIBLIOGRAPHY ON NEXT PAGE
Format 11**

NAME ━━━━━━━━━━━━━━━━━━━━━━━━━━━━━━━━━━━━━ DATE ━━━━━━

Title on
line 13

NOTES ↓3

1. Ronald Davis Franco, <u>Computing Made Simple</u>, Henley
Publishing Company, Inc., New York, 1983, p. 234. ↓2

No
superscripts

2. Linda H. Solomon, "Microcomputing for Offices,"
<u>Automation Trends</u>, November 1982, pp. 43-45.

3. Renaldo S. Gonzales, "Micros for Your Office," <u>Data
Magazine</u>, January 1983, pp. 67-69.

4. Ibid., p. 70.

5. Solomon, p. 46.

Title on
line 13

BIBLIOGRAPHY ↓3

Entries
alphabetized
by author's
last name

Franco, Ronald Davis, <u>Computing Made Simple</u>, Henley Publishing
    Company, Inc., New York, 1983. ↓2

Gonzales, Renaldo S., "Micros for Your Office," <u>Data Magazine</u>,
    January 1983.

Solomon, Linda H., "Microcomputing for Offices," <u>Automation
    Trends</u>, November 1982.

**REPORT: DOCUMENTATION PAGES
ENDNOTES AND BIBLIOGRAPHY
Format 12**

HEADERS AND FOOTERS

Depending on formality and need, business reports may use headers and footers. An identifying label (always the same words) typed at the top of continuing pages is a *header*. An identifying label (always the same words) typed at the bottom of continuing pages is a *footer*.

Headers and footers are usually typed on each page (after the first page) or on every other page. Headers or footers can be used in many ways such as to identify continuous pages of a report, date the pages of price sheets, or number the pages of a policy manual.

Some types of automated equipment have header and footer commands that allow the keyboarder to enter the header or footer once. When the document is printed, the header or footer is printed on each page or on every other page as specified.

Automatic page numbering is usually considered part of the header-footer command. Once the command is entered, the pages are numbered consecutively in the memory, on the screen, and on the printout.

EXERCISE 4-7

In the blank provided, write the word from the list that best completes each numbered item. Words may be used more than once. If you are in doubt about an answer, check the text explanations and Formats 9 to 12.

bibliography	footnotes
by-line	header
dividing line	notes
endnotes	side heading
footer	title

1. List three report parts that begin on line 13, are centered, and are typed in all capitals.

   a. _____

   b. _____

   c. _____

2. Numbered references at the bottom of report pages are _____.

3. Numbered references arranged on one page near the end of the report are

   _____.

4. An alphabetized listing of sources is the _____.

5. An identification label of the same words typed at the top of each page is called a/an

   _____.

6. An identification label of the same words typed at the bottom of each page is called a/an

   _____.

7. Single-space before and double-space after this report part: _____.

8. Double-space before and triple-space after this report part: _____.

NAME ▬▬▬▬▬▬▬▬▬▬▬▬▬▬▬▬▬▬▬▬▬▬▬▬▬▬▬ DATE ▬▬▬▬▬▬▬▬▬

REVIEW EXERCISE 4-A

Use revision symbols to correct format errors and punctuation style errors in the six items that follow.

1. Memorandum on plain paper

MEMORANDUM

To: Liza Berger, Supervisor
           Production Department
From: Maurice Lemaire, Manager
Date: June 20, 198-
Subject: New Stationery

Your order for personalized stationery was received
yesterday. We are ready to begin production on the
500 sheets of standard letter size and 250 note-size
sheets that you requested.

We are pleased to announce that we now have three
attractive new styles of letterhead available. I
have attached a sample of each and a form on which
you should indicate the style you prefer.

Please complete the form and return it to me
immediately. Your order will be ready two weeks from
the date we receive your response.

Thank you for using our services.

kb

NAME ━━━━━━━━━━━━━━━━━━━━━━━━━━━━━━━   DATE ━━━━━━━━━

2. Social-business letter, modified-block style, standard punctuation, and indented paragraphs

3829 Oxford Road
Springfield, Oh 44904

Ms. Phyllis Jamison
8827 Lakecrest Drive
Springfield, OH 44903

Dear Ms. Jamison:

Congratulations on being named Businesswoman
of the Year by the Springfield Chamber of Commerce.
It is a pleasure for those of us who know you per-
sonally to see your contributions recognized by the
entire community.

Doubtless, receiving this prestigious award will
result in your busy schedule becoming even more
hectic. Therefore, I would like to get in an early
invitation for a celebration luncheon with the
officers of the Springfield Small Business Owners
Association.

I will contact your office with a formal
invitation within the next two weeks. Meanwhile,
I could not wait to let you know how happy I was
to hear the good news.

Sincerely yours,

Cathy Raines
Cathy Raines

NAME ━━━━━━━━━━ **DATE** ━━━━━━

3. Simplified letter

Data Company, Inc.

3281 Kenan Drive • Flint, Michigan 48504 • (313) 555-6391

May 10, 198-

Mr. Ronald Lee
690 Bellevue Avenue
Flint, MI 48506

Programming Position Open

It has come to my attention that you are interested
in exploring job opportunities with Data Company, Inc.
We are currently seeking applicants for a position
as Junior Programmer in our order fulfillment
department.

This position offers a competitive salary with flexible
working hours. In addition to on-the-job training, we
offer successful employees at Data Company a full range
of opportunities for professional development.

     If you would like more information on the
challenging position that is now open, please contact
me at the telephone number above. I look forward
to hearing from you.

Frank Luciano

FRANK LUCIANO--PERSONNEL MANAGER

dp

NAME ▰▰▰▰▰▰▰▰▰▰▰▰▰▰▰▰▰▰▰▰▰▰▰▰▰▰▰▰▰   DATE ▰▰▰▰▰

4. Two-page block-style letter, attention line, and open punctuation

4736 Newport Drive
Flint, Michigan 48504
(313) 555-8745

October 21, 198-

Attention: Mr. Joe Bianco

Reynolds Construction
624 Bellevue Avenue
Brockton, MA 02402

Dear Mr. Bianco

Flint Systems, Inc., is planning a major expansion in
several key departments of the company. Plans are
currently being finalized and by early spring we should
be ready to begin renovation on newly acquired office
space to accommodate approximately 30 new employees.

contact you to set up a meeting to discuss specifications and estimated costs of such construction. To
help you in coming up with some preliminary figures,

Reynolds Construction   2   October 21, 198-

I am enclosing a floor plan.

Thank you for your quick response to my inquiries.
Please call me if you have any questions that need
to be answered before our meeting.

Sincerely yours

Anna J. Goldberg

Anna J. Goldberg
Manager, Building Operations

mt
Enclosure

NAME ━━━━━━━━━━━━━━━━━━━━━━━━━━━━━━━━━━ DATE ━━━━━━

5. Original letter, modified block, indented paragraphs, and standard punctuation

MANAGEMENT CONSULTANTS, INC.

3699 FAIRWOOD ROAD · CINCINNATI, OHIO 45239 · (513) 555-6823

June 2, 198-
bcc: Mr. Drew Deaton

Executive Office Services
5601 Dulap Road
Cincinnati, OH 45247

Attention: Ms. Pat Demay

Ladies and Gentlemen

Subject: Electronic Equipment Exhibition

    This letter will confirm plans for the exhibition
of electronic office equipment being sponsored by
Executive Office Services for the employees of Management Consultants, Inc. The exhibition will take place
on June 17, from 9:30 a.m. to 4:00 p.m., in the
auditorium.

    The signed contract for your services and a final
copy of the exhibit program are enclosed.

                Sincerely yours
                MANAGEMENT CONSULTANTS, INC.

                Logan Bolinsky
                Senior Consultant

rk
2 Enclosures
Certified
cc: Mr. Esteban Cordova
    Mrs. Lillian Soo
    PS: If the exhibition is as successful as we
anticipate, we will probably want to repeat it next

year.

NAME ━━━━━━━━━━━━━━━━━━━━━━━━━━━━━ DATE ━━━━━━

6. Page 1 of a two-page report

MICROCOMPUTERS

By Freida Griffen

The purchase of microcomputers for use by executive level staff continues to rise as availability of useful software applications and keyboarding proficiency among executives increase. Throughout the 1980s, the number of executives using their own personal computers in the office is expected to rise.[1]

Equipment

A survey of the purchasing patterns among executives revealed that purchase of a specific brand is most often based on the number and types of software packages available. The survey also revealed some other interesting factors that make a difference.[2]

Brands of Computers. The following list of examples taken from the survey points out the variety of reasons executives have for purchasing a particular brand of computer:

1. Brand A has a "mouse" that moves the cursor and controls interaction with the program, instead of relying solely on a keyboard.

2. Brand B was the first company to offer a free keyboarding instructional package with the purchase of a microcomputer.

3. Brand C is the most easily portable microcomputer.

Printers. Quality of the printed product and speed of printing are two important factors in the selection of printers.

---

[1]Ronald Davis Franco, Computing Made Simple, Henley Publishing Company, Inc., New York, 1983, p. 234.

[2]Linda H. Solomon, "Microcomputing for Offices," Automation Trends, November 1982, pp. 43-45.

REVIEW EXERCISE 4-B

As personnel manager of Country Vehicles, Inc., you approve items going to the newspaper or the printer. To avoid errors, you have all items typed in the desired format. Review the following three jobs. Using editing symbols, insert in the typed copy any obvious details not in the original, handwritten quick note. Correct any errors made by the typist. Initial correct items in the upper right corner.

Job 1

QUICK NOTE

Please approve for printing one dozen "from the desk of" memo pads for each person listed:
Alice Fremont
Dan Trammell
Deite Deveron

Thanks,
Bill Straton

from the desk of
ALICE FREMONT

from the desk of

from the desk of
DIETE DEVERON

Job 2

QUICK NOTE

Please approve this ad for a secretary.

AD

Position with variety. Top skills required.
Experience preferred. Good pay and
fringe benefits. Send letter and résumé
to Personnel Manager, Country Vehicles, Inc.,
2525 Eastman Avenue, Denver, CO 80210

Thanks,
Susan Lee

DRAFT OF NEWSPAPER AD

Position with variety. Top skills
required. Experience preferred. Good
pay and fringe benefits. Send letter
and résumé to Personnel Manager,
Country Vehicles, Inc., 2525 Eastman
Avenue, Denver, CO 80210.

Job 3

QUICK NOTE

Please approve this ad:

SPECIAL SALE
FOUR-WHEEL-DRIVE VEHICLES
Saturday, April 5, 7 a.m. to 7 p.m.
Prices too low to quote.

Thanks!
Kim Mann

DRAFT OF NEWSPAPER AD

SPECIAL SALE

FOUR-WHEEL-DRIVE VEHICLES

Saturday, April 5, 7 a.m. to 7 p.m.

Prices too low to quote!

NAME ▬▬▬▬▬▬▬▬▬▬▬▬▬▬▬▬▬▬▬▬ DATE ▬▬▬▬

IS IT CLEAR?

CHAPTER

Effective Use of Words

Choose the Right Words

All writing—especially business communication—must be clear. These are comments that you *don't* want:

"What does this word mean?"
"This doesn't make sense."
"I can't figure it out!"
"It sounds good, but what does it say?"
"It's as clear as mud. Right?"

Use these techniques to make messages clear:

Use simple words.
Use proper English.
Avoid trite language.
Spell correctly.

Use Simple Words

Notice how simple words can replace harder words without changing the meaning. Both sentences are correct, but the second one is easier to understand.

1. O'Neil will *ascertain* the facts.
   O'Neil will *get* the facts.
2. You should be *cognizant* of our policies.
   You should be *aware* of our policies.
3. The program will *be implemented* today.
   The program will *begin* today.
4. *Utilize* the best method.
   Use the best method.
5. She is *contemplating* the matter.
   She is *thinking about* the matter.

Use Proper English

The first sentence in each pair uses substandard English. The second sentence corrects the error.

1. **Wrong**   We interviewed *alot* of applicants.
   Right   We interviewed *a lot* of applicants.
2. **Wrong**   Mail is delivered *irregardless* of the weather.
   Right   Mail is delivered *regardless* of the weather.
3. **Wrong**   Steve will *try and hire* six people.
   Right   Steve will *try to hire* six people.
4. **Wrong**   Martha *should of* gone.
   Right   Martha *should have* gone.
5. **Wrong**   His answers were *alright*.
   Right   His answers were *all right*.

NAME ▬▬▬▬▬▬▬▬▬▬▬▬▬▬▬▬▬▬▬▬▬▬ DATE ▬▬▬▬

EXERCISE 5-1

Change hard words to easier words and correct word usage errors. Replace each underlined word or phrase in the memo draft with one of the words or phrases listed below. Use revision symbols to show the changes.

ask for	resulting
begin	improving
help	try to find
regarding	

MEMORANDUM

To: Al Poleta
From: Ken Stacy
Date: March 18, 198-
Subject: Telephone Log

In regards to your suggestion, Kaye Haskins, of
Consultants, Inc., will initiate a study to try
and find ways of facilitating our communications.
She will call you tomorrow.

Please explain the study to your staff and solicit
their assistance. They should be involved in the
study and the subsequent decision.

KS

HOMONYMS

Homonyms are words that sound alike but are spelled differently. For example, the words *to, too,* and *two* are all homonyms. Two groups of homonyms are shown below.

Increasing the speed of the machines may *affect* the quality.
What *effect* will the decision have?

I appreciate your *advice.*
The lawyer *advised* her client to sell the property.

EXERCISE 5-2

Provide definitions for each of the words below. Note that the part of speech (for example, noun or verb) may change the definition. Include the part of speech for each of these words.

1. affect _____

_____

2. effect _____

_____

3. advice _____

_____

4. advise _____

_____

EXERCISE 5-3

Are the underlined words in the memo draft used correctly? Cross through an incorrectly used word, and write the correct word above it. Use these words for any changes.

accept, except	approved, improved	since, sense
advice, advise	know, no	their, there
affects, effects	sale, sell	to, two, too

MEMORANDUM

To: Jon Shalsky, Sales Representative
From: Cathryn Peeler, Sales Manager
Date: August 25, 198-
Subject: Spartan Motel Order, No. 3216

Congratulations, Jon, on getting the large drapery order from the Spartan chain. This is your biggest <u>sell</u> <u>sense</u> joining us.

As you <u>no</u>, a large order <u>effects</u> many departments. Our credit department has <u>improved</u> Spartan's credit request, allowing us to <u>except</u> the order. The planning department scheduled the work <u>too</u> meet Spartan's needs. The printing and shipping departments have adjusted <u>there</u> schedules.

```
        Mr. Spartan will be here during the dyeing and
        printing process to advise us on acceptable colors and
        quality.  I would appreciate your working with him
        October 20-25.

        Keep up the good work!

                        CP
```

ONE WORD OR TWO?

Depending on usage, some words can be written together in some sentences and written separately in others. Here are several frequently occurring examples.

Spell *anyone, everyone,* and *nobody* as one word unless they are followed by *of.*

1. *Any one* of the applicants can fill the job.
   Anyone can fill the job.
2. *Every one* of them should be promoted.
   Everyone should be promoted.
3. *No body* of people has that much influence.
   Nobody has that much influence.

Here are some other pairs that are frequently misused.

    already, all ready
    altogether, all together
    always, all ways
    anyway, any way
    indirect, in direct
    maybe, may be
    sometime, some time

Always check a dictionary or a reference book when you are unsure whether one word or two words should be used. The *Gregg Reference Manual* (William A. Sabin, *The Gregg Reference Manual*, 6th ed., McGraw-Hill Book Company, New York, 1985) clarifies some of these questions.

EXERCISE 5-4

Use revision symbols to correct errors (if any) in the italicized words.

1. *Everyone* worked late.

2. Our records are *all ready* for the audit.

3. Our method is *indirect* conflict with company policy.

4. *No body* wanted the job.

5. Does *any one* have a suggestion?

6. Put the color samples *all together* for comparison.

7. *All ways* check quality.

NAME ▬▬▬▬▬▬▬▬▬▬▬▬▬▬▬▬▬▬▬▬▬▬▬▬ DATE ▬▬▬▬▬▬▬▬

8. Buying new equipment *maybe* the answer.

9. We need *sometime* to make a decision.

10. Help the new employee in *any way* you can.

Avoid Trite Language

Business jargon that has lost its effectiveness through overuse is trite. These expressions often seem "empty" as a result of wordiness. Trite expressions can be improved.

1. **NO**   Enclosed please find your new watch. (Is it lost?)

   YES   Your new watch is enclosed.

2. **NO**   We are in receipt of your request for samples and have enclosed them herewith. (Stiff, impersonal, and wordy)

   YES   Here are the samples you requested.

3. **NO**   We would be favored with a written reply. (Wordy and obsolete)

   YES   Please write us.

4. **NO**   The directories will be sent under separate cover. (Trite and obsolete)

   YES   The directories will be sent separately.

5. **NO**   As per your request for a catalog, we have enclosed same. (Stiff and wordy)

   YES   The catalog you requested is enclosed.

Participial closings are considered obsolete and trite. These closings are sentence fragments that begin with an *-ing* word.

6. **NO**
Thanking you for your order, I remain,
> Sincerely yours,
> J. B. Bryson

YES
Thank you for your order.
> Sincerely yours,
> J. B. Bryson

7. **NO**
Expecting your check soon, I remain, . . .

Yes
Please send us your check promptly.

8. **NO**
Looking forward to meeting you, I am, . . .

YES
I look forward to meeting you.

NAME ━━━━━━━━━━━━━━━━━━━━━━━━━━━━━━━━━━━━ **DATE** ━━━━━

EXERCISE 5-5

Using revision symbols, edit this letter draft to remove trite expressions. Make sure your edited version includes these facts: Arnold Carson hopes he helped Ms. DiStefano by returning the questionnaire she enclosed with her November 1 letter. The completed questionnaire is enclosed.

PORTLAND frozen food, inc.
6405 Dolph Street, S.W. • Portland, Oregon 97219 • (503) 555-2837

November 23, 198-

Ms. Susan DiStefano
2384 Holly Lane
Portland, Oregon 97223

Dear Ms. DiStefano:

We are in receipt of yours of the first asking us to complete a questionnaire. Please find same completed and enclosed herewith.

Trusting we were helpful in your studies, we remain,

Sincerely yours,

Arnold Carson

jk
Enclosure

Spell Correctly

Look up unfamiliar words or words that don't look quite right.

Some spelling errors are simply misspellings where no particular rule applies. Other spelling errors occur because rules are not applied. Often used rules are listed in the Reference Section. Learning these rules can reduce errors.

When in doubt, find out! Use a dictionary or a wordbook. A wordbook doesn't give as much information as a dictionary does, but a wordbook is a quick, convenient source for checking spelling or syllabication.

Shown here is page 175 from *20,000 Words* (Louis A. Leslie, *20,000 Words*, 7th ed., McGraw-Hill Book Company, New York, 1977). Use it to correct the spelling errors in Exercise 5-6.

re·bus	re·cline	re·cru·des·cence
re·but·tal	re·cluse	re·cruit
re·cal·ci·tran·cy	rec·og·ni·tion	rect·an·gle
re·cal·ci·trant	rec·og·niz·able	rect·an·gu·lar
re·cant	re·cog·ni·zance	rec·ti·fi·ca·tion
re·ca·pit·u·late	rec·og·nize	rec·ti·fi·er
re·ca·pit·u·la·tion	re·coil	rec·ti·fy
re·cap·ture	re—col·lect	rec·ti·lin·ear
re·cede	(collect again)	rec·ti·tude
re·ceipt	rec·ol·lect (recall)	rec·to·ry
re·ceiv·able	rec·ol·lec·tion	re·cum·bent
re·ceive	rec·om·mend	re·cu·per·ate
re·ceiv·er·ship	rec·om·men·da-	re·cu·per·a·tion
re·cent	tion	re·cur
re·cep·ta·cle	re·com·mit	re·curred
re·cep·tion	rec·om·pense	re·cur·rence
re·cep·tive	rec·on·cile	re·cur·rent
re·cess	rec·on·cil·i·a·tion	re·cur·ring
re·ces·sion	re·con·di·tion	red·bird
re·ces·sion·al	re·con·firm	red—blood·ed
re·ces·sive	re·con·fir·ma·tion	red·breast
re·cid·i·vism	re·con·nais·sance	red·bud
rec·i·pe	re·con·noi·ter	red·cap
re·cip·i·ent	re·con·sid·er	red—car·pet adj.
re·cip·ro·cal	re·con·struc·tion	red·coat
re·cip·ro·cate	re·cord v.	Red Cross
re·cip·ro·ca·tion	rec·ord n.	re·dec·o·rate
rec·i·proc·i·ty	re·cord·er	re·deem
re·cit·al	re·course	re·deem·able
rec·i·ta·tion	re·cov·er (regain)	re·deem·er
rec·i·ta·tive	re—cov·er	re·demp·tion
re·cite	(cover again)	re·de·vel·op·ment
reck·less	re·cov·ery	red—hand·ed
reck·on	rec·re·ant	adj., adv.
re·claim	rec·re·ation	
rec·la·ma·tion	re·crim·i·na·tion	

EXERCISE 5-6

These five errors occur in the paragraph below. Use revision symbols to make the corrections.

1. *A word division error. (Records is a noun.)*
2. *An ei error.*
3. *Error with consonant before suffix.*
4. *A simple misspelling.*
5. *A simple misspelling.*

Tom found the reason for the recuring problem. We do not keep re-cords of telephone adjustment requests. Replacements are made as soon as the defective items are recieved. After recognising this problem, Tom made these reccomendations.

EXERCISE 5-7

Edit the following for spelling errors. Use revision symbols to show corrections. Check any spelling that doesn't look right. The Reference Section on spelling can help. (Note that items may be correct or have more than one error.)

1. Beth, your real estate agent, wrote a letter of introduction to your new nieghbors.

2. Your check will be your reciept.

3. The broker has recieved your stock certificates.

4. He implied that I would be promoted.

5. Managment feels that all imployees are important.

6. Dr. Jose Sartina refered the pateint to you.

7. My account executive reccomends buying bonds.

8. Mail your application to Amy Scarelli, Personel Manger.

9. Betty's performance and professional achievment convinced us to hire and promote more handicapped people.

10. Planing is a key to success.

11. Continueing the project will reduce useable materials.

12. We had sales totalling $500 the first week.

13. They are tieing ties for the sales display.

14. Under flexable scheduleing, you choose your working hours.

15. Believeing in the product helps you sell it.

NAME _____ DATE _____

SPELLING VERIFICATION

Some word processing software finds spelling errors. Words that aren't "recognized" by the equipment are highlighted. The operator must decide if the word is correct or incorrect.

Proofreading software has a basic dictionary of words that it recognizes. Operators can add words that are peculiar to their organizations. For example, new medical terms or chemical terms that would not be in the software dictionary can be added so that the software will recognize them and not highlight them as possible errors.

However, software that proofreads for spelling errors can lead to a false sense of security. Look at this example.

1. Sentence appearing on the screen:

    We recieved tow new brooks form him.

2. Intended sentence:

    We received two new books from him.

3. Errors highlighted by software:

    We *recieved* tow new brooks form him.

4. *Can you underline four errors in the sentence?*

    We recieved tow new brooks form him.

5. *Compare the number of errors found.*

    Software found _____ error/s.

    You found _____ error/s.

6. *Did you find these errors?*

    We *recieved tow* new *brooks form* him.

Software highlights letters that don't form words. A misspelling that results in a correctly spelled word, even though it's the *wrong* word, will not be detected as a possible misspelling by the software. Even with automated equipment, quality depends on editing skill!

WORD DIVISION

Dividing words correctly makes messages easier to understand. Use a dictionary or a wordbook for syllabication. Refer to the rules given in the Reference Section for word division.

EXERCISE 5-8

Use revision marks to edit for correct word division. If the item is incorrectly divided, move the letters. Move or delete the hyphen as needed. To stay within specified document margins, you can add only one *or* two *strokes to the right margin. The hyphen is considered one stroke. Divide items at the* preferred *point.*

Examples

a.   Xxxxxx xxxxxxx xxx

b. xxxxx xxxx xxxxxx xxxxx (li)

c. nes. Xxxxx xxxxx xxx xxxx

d. xxxx xxxxxx xxxx xxxxx subd-

NAME _____   DATE _____

e. ivision xxx xxxxxx xxxxx 1,000,

f. (276)xxx xxxxx xx xxxxx xxxx xx

g. xxxxxxx xxxx xx xxxxxx. Xx

h. xxxxx. X xxxxx xxxx xxx $93.

i. (24)xx xxxx xxxxxx xxxxx xxxxx.

Line b: *Lines* has only one syllable and shouldn't be divided. The margin will not allow moving four strokes (*nes.*) from line c to line b.
Line d: *Sub-di-vi-sion* has three syllable breaks, but the preferred division point is after the prefix *sub*.
Line f: Move *276* to the previous line. Deleting the hyphen stroke will allow moving three strokes (*276*) to line e.
Line i: Move *24* to line h. The margin allows moving two strokes.

1. Yesterday, we received a letter from Henson Company about our August seminar. They want to send eight people.

2. All of the members agree on the issue. The vote will affirm our position for a formal statement to the media.

3. To meet our specific company needs, the programmer will ideally know these three languages: COBOL, BASIC, and FORTRAN. Do we have applicants that know all three?

4. The durability of our product is achieved by treating and compressing the cardboard. This process is expensive.

5. Dependable employees make my job much easier. Often, I have to rely completely on one of them to close and lock the restaurant.

6. During the last shift, we found that the machine had completely stopped sealing the packages. We had to do this by hand.

7. He became vice president of the corporation when he was almost twenty-two. Most of the employees were glad that he was promoted.

8. Cathy and William have asked their staff to call immediately anytime there is a problem with the equipment.

9. Their latest survey has been completed. They interviewed 1,296, 235 people to find the answers to several health problems.

10. We can use Mrs. Miller's talents with the sales campaign. She is creative, and she expresses her ideas very well.

11. People enjoy working puzzles in their spare time. They like guessing the answers. They search for freedom when doing a maze puzzle. Scrambled words offer an additional challenge to puzzle enthusiasts.

12. Mary L. Leer will do well with the project. She is perceptive and understands the situation. She is dependable and is well qualified to evaluate all aspects.

NAME ——— DATE ———

13.   Zoo keepers have to observe each animal. They noticed that three apes behaved strange-ly after eating the new food. We very candid-ly, but tactfully, wrote the supplier of the new food preparation.

14. **Note:** This is the last paragraph on page 1 of a two-page letter. It is followed by the top of the second page.

   The new plant should be located near an interstate highway and, if at all possible, some-

15.   Sam will be here tomorrow morning at 9:30 to install the new telephone equipment. He could-n't come today because of the bad weather.

Mr. Stan Ward   2   November 1, 198–

where near the railroad. This should help move materials and finished products more efficiently.

   Some word processing software can find possible hyphenation points. After keyboarding a memo, for example, the operator can use the hyphenation command to stop the cursor at each point where word division would make the right margin more even. The **cursor** is a blinking white light on the screen showing the point in the text where the operator is working.

   The operator decides *whether* and exactly *where* to hyphenate. Here are some examples.

   Suppose the cursor showed that *rely* could be divided after the *r* or the *e*. The operator would not divide this word because it is too short but would ask for the next possible hyphenation point.

   Suppose the cursor showed that *recommendation* could be divided at any point up to, and including, the second *e*. The operator would look at the syllable breaks.

   rec-om-men-da-tion

   Then the operator would put as many letters as possible before the hyphen *without* breaking word division rules. In this case, the hyphen should go between the two *m's*.

   Word processing programs from different companies offer various hyphen-ation capabilities. Some software will require more operator skill than others. Exercise 5-9 asks for hyphenation decisions similar to those made on many software programs.

EXERCISE 5-9

In the words on the next page, the large space shows where the cursor in a hyphenation program has stopped. You can divide the word at any point before this space. Insert a hyphen at the preferred division point. Close up the space in words that should not be divided. Remember to get as many letters as possible before the hyphen without breaking word division rules.

NAME ━━━━━━━━━━━━━━━━━━━━━━━━━━━━━━━━━━━━ DATE ━━━━━━━

Examples

gues‿sing
blessi ng

1. beginn ing
2. shouldn 't
3. a warded
4. thir ty-two
5. anybo dy
6. antib iotic
7. antiq ue
8. pref erred
9. pref erable
10. evaluati ve

PLURALS AND POSSESSIVES

Effective use of words includes forming plurals and possessives correctly. Exercise 5-10 offers practice in editing for errors with plurals and possessives. When you are not *absolutely* sure, check the Reference Section.

EXERCISE 5-10

Edit these items for correct plurals and possessives. Use revision symbols to show needed changes. (Review the Reference Section if necessary.)

1. Our company and two other companys donated blankets to disaster victims.

2. Men's benefits and women's benefits are equal.

3. All replies will be kept confidential.

4. Sandford Pressley, Bill Davis's friend, won the contest.

5. Buy one decoy for $12, or buy two decoies for only $20.

6. Major policies were reviewed by both chiefs of staff.

7. Three boxs were damaged.

8. Childrens clothes and toys will be reduced for the sale.

9. The invoices are on my secretarys desk.

10. Cash prizes will be given to two runners-up.

11. Both managers' opinions were considered.

12. John's and Fred's homes are insured. (Separate ownership; each has a home.)

13. John's and Fred's plan worked. (Joint ownership of one plan.)

14. Five employees goals included increasing productivity.

15. All editors in chiefs will attend the budget meeting.

16. Susan called Jim to insist on him coming today.

17. Ann and Tom's home is in the suburbs. (Joint ownership of one home.)

18. Ray and Sue's cars are insured. (Separate ownership; each has a car.)

19. My last customer bought seven rose bushes'.

20. Quick reaction resulted in his saving a life.

NAME ▬▬▬▬▬▬▬▬▬▬▬▬▬▬▬▬▬▬▬▬▬▬▬▬▬▬▬▬▬▬▬▬▬ DATE ▬▬▬▬

REVIEW EXERCISE 5-A

Assume that this memorandum is on your word processing screen. The proof-reading software for spelling does not recognize the highlighted (boxed) words. Those words have been placed on the checklist marked "verify." Add to the list all additional incorrect words and any other words that you question. Using a reference source, verify each one. On the list, strike through the error and write the correction beside it. Put a check (√) beside any that you verify as correct.

```
                    MEMORANDMM

To:       All Sales Personal
Form:     Anthony Peabody
Date:     April 18, 198-
Subject:  Procedure for Sales Returns

The new  proceedure  for sales returns goes into
effect on Monday, May 1.  Thanks for suggesting
this improvement.

The attached explanation should cover all most every
possible situation.  Please  reat  it before our next
meeting.  This procedure effects every ones job
directly or in directly, and we want to make sure
every thing is right.

                    AP

dt
Enclosure
```

Verify

MEMORANDMM
proceedure
reat

REVIEW EXERCISE 5-B

Use revision symbols to edit this two-page letter. Use global symbols whenever possible.

Baxter's
department store

1313 Adams Street
Annapolis, Maryland 21403
(301) 555-9861

May 20, 198-

Mr. Rajan Mitra, P.C.
2856 Leander Street
Annapolis, MD 21403

Dear Mr. Mitra:

Your request for a Baxter's credit card has
been approved. Your account is opened, and your
credit card is enclosed. Unlike many credit cards,
the Baxter's card can help you save money!

Did you notice the <u>P.C.</u> beside your name? Being a
credit customer automatically puts you on our prefered customer list. As a prefered customer, you will be invited to Pre-
Sale Sales. Your card admits you two hours early for all clear-
ance sales. Prefered customers also recieve coupons with their
statements, which give an even greater reduction on sale
items.

Please sign the credit card now. Let us know
at once if the card is lost or stolen. You are not
responsible for any misuse after this notification.

The closing date of your statement is the 15th
of each month. Use your card <u>interest-free</u> by paying the balance by the 1st of the month. If you
prefer, you may make a partial payment of 10 per cent

Mr. Rajan Mitra   2   May 20, 198-

of the balance. Interest on the remaining balance
will be at the rate of 1 1/2 per cent per month (18
per cent per year).

    Mr. Mitra, we are sure that you will enjoy the
convenience and the savings offered by your credit
account. Use your card and the coupon given below
to save an additional 10 percent on all ready reduced
swim wear during June.

    Looking forward to serving you, we remain,

                    Sincerely yours,

                    Walter K. Schmidt

glf
Enclosure


```
        10%          Coupon          10%
                      for
                    SWIMWEAR

             Charged During June 198-
             BAXTER'S DEPARTMENT STORE
        10%          Coupon          10%
```

IS IT COURTEOUS?

CHAPTER

Reader's Viewpoint

Make a Good Impression

Many communications cross the desks of business workers each day. Does that mean that every word of each letter, memo, or report is read? Definitely not! Workers who know how to manage their time will discard some communications after one glance, quickly scan others, and read a select few very thoroughly.

Three qualities can help boost communications into the select group that are thoroughly read. Edit messages to make them:

Eye-pleasing
Reader-centered
Positive

Eye-Pleasing Messages

An attractive, uncluttered, easy-to-read format can help business messages make a good first impression. Corrections should be made neatly, and correct formats (discussed in Chapter 4) should be used. Here are some other techniques for improving the appearance of a message.

ADJUST PARAGRAPH LENGTH

Message content must determine paragraphing decisions. However, when it is practical, adjust paragraph length to fit the guidelines below. Remember, these are guidelines, *not* hard and fast rules.

1. Keep the first and last paragraphs short, usually two to five lines.
2. Have middle paragraphs average four to eight lines and make them longer than the first and last paragraphs.
3. Vary paragraph length.
4. Use several short paragraphs instead of one long paragraph to create an uncluttered appearance.
5. Combine many short paragraphs to correct a choppy appearance.
6. Avoid a heavy, top-heavy, or a bottom-heavy appearance.
   Heavy: All or most paragraphs are too long.
   Top-heavy: Beginning paragraphs are too long.
   Bottom-heavy: Ending paragraphs are too long.

EXERCISE 6-1

Diagnose the paragraphing problems, and prescribe their treatment. Write the letter of the correct diagnosis and the letter of the correct prescription in the blanks provided. If no paragraphing problems exist, use diagnosis c and prescription z.

Diagnosis

a. Heavy, top-heavy, or bottom-heavy appearance.
b. Middle paragraphs are too short.
c. No paragraphing problems.

Prescription

x. Divide long paragraphs.
y. Combine short paragraphs.
z. No corrections needed.

NAME ━━━ DATE ━━━━━━

1. Diagnosis _____   Prescription _____

Dear Mr. Bell:

_____
_____
_____
_____
_____
_____
_____
_____
_____

Sincerely,

2. Diagnosis _____   Prescription _____

Dear Ms. Winn:

_____
_____
_____
_____
_____
_____
_____
_____

Sincerely,

3. Diagnosis _____   Prescription _____

Dear Dr. Dactyl:

_____
_____
_____
_____
_____
_____
_____
_____
_____

Sincerely,

4. Diagnosis _____   Prescription _____

Dear Mr. Omerjee:

_____
_____
_____
_____
_____
_____
_____
_____

Sincerely,

5. Diagnosis _____   Prescription _____

Dear Mrs. Nagamatsu:

_____
_____
_____
_____
_____
_____
_____
_____
_____

Sincerely,

6. Diagnosis _____   Prescription _____

Dear Mr. Diaz:

_____
_____
_____
_____
_____
_____
_____
_____

Sincerely,

NAME ▬▬▬▬▬▬▬▬▬▬▬▬▬▬▬▬   **DATE** ▬▬▬▬

7. Diagnosis _____ Prescription _____

Dear Mrs. Gearren:

_____
_____
_____

_____
_____
_____
_____
_____
_____
_____
_____
_____
_____
_____
_____

Sincerely,

8. Diagnosis _____ Prescription _____

Dear Ms. Van Der Linden:

_____
_____
_____
_____

_____
_____
_____
_____
_____
_____
_____
_____
_____
_____
_____

Sincerely,

9. Diagnosis _____ Prescription _____

Dear Miss Friedman:

_____
_____

_____
_____
_____
_____

_____
_____
_____
_____
_____
_____
_____

Sincerely,

10. Diagnosis _____ Prescription _____

Dear Professor Wu:

_____
_____
_____
_____
_____

_____
_____
_____
_____
_____
_____
_____
_____

Sincerely,

NAME _____ **DATE** _____

USE COLUMN OR ENUMERATED FORMATS

Combining columns or enumerations with a paragraph format improves the appearance of a message and makes it easier to read.

COLUMN FORMAT

Letters A and B show the same part of a social-business letter in two formats—paragraph format (A) and column format (B). The column format is much easier to read. Notice the extra line of space separating the first two lines from the last two lines. This is an especially good way to treat a column with numbers.

The revision symbols in letter A tell the typist to use a column format, to add headings, and to insert an extra line of space.

Letter A: Paragraph Format

Please send me the following: 2 pocket calculators, catalog no. CZ-6132, at $10 each (Total $20); 2 power adapters, catalog no. CZ-6133, at $5 each (Total $10); 2 battery chargers, catalog no. DF-6787, at $10 each (Total $20); and 5 pens, catalog no. GF-2847, at $5 each (Total $25). The total for all items is $75.

[A] Change to column format. Headings: line space Quantity Description Catalog Number Price Total

Letter B: Column Format

Please send me the following:

Quantity	Description	Catalog Number	Price	Total
2	Pocket Calculators	CZ-6132	$10	$20
2	Power Adapters	CZ-6133	5	10
2	Battery Chargers	DF-6787	10	20
5	Pens	GF-2847	5	25
			Total	$75

ENUMERATED FORMAT

Memos A and B show the same part of a memo in two formats—paragraph format (A) and enumerated format (B). The enumerated format is much easier to read.

Revision symbols in Memo A tell the typist to indent the block five spaces from each margin and enumerate the items.

Memo A: Paragraph Format

There are three reasons that we need an additional office employee by June. Increased sales have created more paperwork. Our billing is behind, which causes cash flow problems. Operating this summer, when most employees take vacation, will delay billing and other procedures.

5 [B] Enumerate reasons 5

Memo B: Enumerated Format

There are three reasons that we need an additional office employee by June.

1. Increased sales have created more paperwork.

2. Our billing is behind, which causes cash flow problems.

3. Operating this summer, when most employees take vacation, will delay billing and other procedures.

NAME ━━━━━━━━━━━━━━━━━━━━━━━ DATE ━━━━━━━

EXERCISE 6-2

Combine columns or enumerations with paragraphs to improve the appearance of these two memorandums and make them easier to read. On the drafts, mark blocks and use marginal notes to show the changes. Your instructor may ask you to type the memos in the edited format on separate sheets of paper, making sure to capitalize the first letter of each enumeration.

MEMORANDUM

To: Ann Gray-Foster, Records Department
From: Walter K. Wilson, Personnel Department
Date: November 1, 198-
Subject: New Employees

As you requested, we have listed respectively the name, social security number, position number, and date of employment for each new employee. Angela Bates Brunswick, 268-41-1961, 11-27-892, 10/11/8-; Joseph E. Franella, 238-63-8878, 12-36-885, 10/16/8-; Peter H. Petrelli, 261-55-8101, 12-18-187, 10/16/8-; Anna S. Rosen, 285-22-7798, 14-22-772, 10/21/8-; J. B. Sandford, 221-66-5523, 16-22-645, 10/27/8-; and Dana Allen Wong, 331-76-1325, 17-33-474, 10/30/8-. Complete personnel records will be sent to you by November 10.

                              WKW
    ch

MEMORANDUM

To: Walter K. Wilson, Personnel Department
From: Marian Carillo, Vice President
Date: November 3, 198-
Subject: Fringe Benefits

Walter, the executive committee of our board of directors is considering adding one or more benefits. By December 15, will you give us a report on the cost of adding the following: dental insurance for all employees, a physical fitness program for all employees, major medical coverage for all employees, annual physicals for supervisors, and salary continuation insurance for supervisors. Please call if you have questions.

                              MC
    rd

NAME ━━━━━━━   DATE ━━━

Reader-Centered Messages

Stimulate interest by keeping the spotlight on the reader—not the writer—and by personalizing the message.

Spotlighting the Writer

I want to say that *I* think *you* did a great job.

In this example, both references to the writer came before the only reference to the reader. The references to the writer make the sentence wordy. Try to move the *you* to the beginning of the sentence.

Spotlighting the Reader

You did a great job.

The revised sentence above spotlights the reader and is concise.

Spotlighting the Reader and Personalizing the Letter

Sam, you did a great job.

Addressing the reader personalizes the message. Names can be used at the beginning or the end of a sentence and within a sentence. Notice that names used in direct address are separated from the rest of the sentence by commas.

At the Beginning of a Sentence

Frank, do you think the item will sell?

Within a Sentence

Do you think, Frank, that the item will sell?

At the End of a Sentence

Do you think that the item will sell, Frank?

EXERCISE 6-3

Follow the directions to edit the sentences. Use revision symbols to mark changes. Remember to mark punctuation and capitalization resulting from the changes. Then, write the new sentence.

1. Delete three words to make the sentence reader-spotlighted.

   I want to thank you for the generous donation.

   _____

2. Personalize by adding *Alice* at the beginning of the sentence.

   What is your opinion?

   _____

3. Personalize by adding *Mr. Freeman* within the sentence.

   Do you think that the bid is too high?

   _____

4. Personalize by adding *Jim* at the end of the sentence.

   Will you meet me in Dallas?

   _____

NAME ■■■■■■■■■■■■■■■■■■■■■■■■■■   DATE ■■■■■■■■■■■

5. Delete words to make this sentence reader-spotlighted.

I feel that I should tell you that you did a great job.

_____

Positive Messages

Positive communications convey a good feeling and attitude to the reader. These suggestions will make messages more positive.

1. Use positive, rather than negative, words like these: *don't, can't,* and *not.*
2. Use tactful language, and do not insult the reader.
3. Use active rather than passive voice.

   Active voice—The subject is the "doer" of the action:
   Ryan mailed the check.

   Passive voice— The subject is the "receiver" of the action:
   The check was mailed by Ryan.

EXERCISE 6-4

Follow the directions to edit these sentences. Use revision symbols to mark changes. Remember to mark punctuation and capitalization resulting from the changes. Then, write the new sentences.

1. Delete three words to make the sentence more positive.

   If you have any questions, please don't hesitate to call me.

   _____

   _____

2. Delete one word to remove the insult.

   The computer is user friendly. Even you can learn to operate it quickly.

   _____

   _____

3. Remove tactless, unneeded words.

   I am too busy and cannot meet with you before Friday. Can you come to my office at 10:30 a.m.?

   _____

   _____

4. Delete the negative comment.

   I can't finish by Friday, but I will have the report on your desk early Monday morning.

   _____

   _____

5. Change from passive to active voice.

   Your promotion and raise were earned by you.

   _____

   _____

NAME ■■■■■■■■■■■■■■■■■■■■■■■■   DATE ■■■■■■■■

REVIEW EXERCISE 6-A

Use revision symbols to edit the memorandum draft. Rewording will be necessary. Your instructor may ask you to type the edited memo on a separate sheet of paper.

1. Spotlight the reader. Personalize by using the reader's first name at the beginning of this sentence: I was impressed. . . .
2. Remove insults and change negative wording.
3. Change the third and fourth paragraphs to active voice.
4. Combine three short paragraphs.

MEMORANDUM

To: Larry Alexander
From: Fran Lorrance
Date: April 22, 198-
Subject: Compensation Report

I want to thank you for sending me your report on improving our compensation package. You carelessly failed to send me four pages. Please send me pages 35-38, which were missing from my copy.

I was impressed with but surprised by the thorough job you did.

A commendation letter was written to you today by our regional manager.

A copy of the letter has been placed in your personnel file.

Can we discuss the report in my office Monday morning, May 2, at 10:30? I'm sorry that I don't have time before then.

FL

jl

REVIEW EXERCISE 6-B

Making action easy for your reader is another way to be courteous. Coupon letters make responding easy. Use revision symbols to edit the coupon letter on the next page. Your instructor may ask you to type the edited letter on a separate sheet of paper.

1. Edit for errors in mechanics such as capitalization, number usage, formation of possessives, spelling, word usage, and typography (typos).
2. Remove negative comments and insults.
3. Make five (instead of two) paragraphs.
4. Personalize the letter by using the reader's name.

NAME ■■■■■■■■■■■■■■■■■■■■■■■ DATE ■■■■■■■

Medical Software Specialists

Post Office Box 2818
Columbus, Georgia 31902
(404) 555-3868

February 15, 198-

Dr. Lorenzo Tassitino
4826 mitchell Road
Columbus, GA 31907

Dear Dr. Tassitino:

Have you heard of MED-RECORD software for accounting and patient records? This software is for microcomputers and is designed specifically for medical offices. Dr. Lane Tipton, one of your colleagues, bought the program six month's ago. She and her office manger, Don Rosen, are so convinced that MED-RECORD has saved time and money that they are letting us use there names as satisfied customers.

You have to see it to believe how simple it is to use. Your staff members, if they can think and type at all, can quickly learn this software. Our instructor will teach them in your office at the time you you request. I will be in your area March 7 and March 8. Let me demonstrate how MED-RECORD can make your invoiceing and your patient record system more efficient. This will not be a high-pressured sales presentation; MED-RECORD sales itself. Just complete the form below and return it in the enclosed, ready-to-mail envelope. If you don't hesitate but respond before February 25th, you will earn a 10 percent discount on your purchase.

Sincerely,

Anthony Romero

scc
Enclosure

_____   _____
    Name of Practice   Your Name and Title

_____   _____
    Street Address   Brand of Microcomputer

_____   March 7   or   March 8
City   State   ZIP   Preferred Date--Circle One

_____   a.m./p.m.
    Phone Number   Preferred Time

CHAPTER

7

Words

Too Many Words

Wordiness is the use of words that add little or nothing to the sentence. If you can omit words, phrases, sentences, or paragraphs without changing the meaning, the message is wordy. Wordy messages decrease effectiveness and often lead to confusion.

Most business people receive a lot of mail each day. Their demanding schedules limit the time they can spend reading correspondence. They want to take appropriate action without taking valuable time to read unnecessary words.

The exercises in this chapter will give you practice in finding and removing unnecessary words.

EXERCISE 7-1

Strike through one *unnecessary word in each sentence.*

1. Type up the letter.

2. Where is he moving to?

3. The reduced items were sold out in an hour.

4. What time do you need the report before?

5. Where is my pen at?

6. Sort out the mail by ZIP Code.

7. The sales brochure must be redone over.

8. Communication skills are more important than ever before.

9. Our board meeting starts promptly at about 10:30 a.m.

10. We closed up the store early.

EXERCISE 7-2

Strike through five words in this memo draft that you can omit without changing the meaning or causing errors.

MEMORANDUM

```
To:       John Randall
From:     Carmen Ruiz
Date:     August 5, 198-
Subject:  Delivery of Word Processing Systems


Sally Kilminster, with Word Processing, Etc., will
deliver and connect up our two new word processing
systems at about 9:30 Monday morning.  I'll call
you up when she arrives here.

Which system do you want?  They are both alike
except for the color.  One is black, and one is gray.
```

                        CR

        kl

NAME ▬▬▬▬▬▬▬▬▬▬▬▬▬▬▬▬▬▬▬▬▬▬▬▬▬ **DATE** ▬▬▬▬▬▬

EXERCISE 7-3

The numbered items are examples of wordy language. For each numbered item, write the letter of the concise item that has the same meaning.

a. usually   f. weekly
b. helped   g. while
c. if   h. because
d. now   i. always
e. before   j. soon

_____   1. at the present time

_____   2. in the event that

_____   3. due to the fact that

_____   4. every seven days

_____   5. in almost every instance

_____   6. gave assistance to

_____   7. in the near future

_____   8. during the time that

_____   9. at all times

_____   10. prior to

EXERCISE 7-4

The numbered sentences are examples of wordy language. For each numbered sentence, write the letter of the concise item that has the same meaning. Using revision symbols, revise the wordy sentences. Then write the new sentence.

a. annually   d. believes
b. each December e. reported
c. for

_____   1. We set our goals in the last month of each year.

_____

_____   2. Our check in the amount of $24.98 is enclosed.

_____

_____   3. Kim is of the opinion that the plan will work.

_____

_____   4. Hansen gave a report to the effect that the quality is better.

_____

_____   5. Insurance premiums are paid once each year.

_____

NAME ▬▬▬▬▬▬▬▬▬▬▬▬▬▬▬▬▬▬▬▬ **DATE** ▬▬▬▬▬

Redundancy

Redundancies are needless repetitions that express the same thought. Redundant words are very close in meaning and say essentially the same thing.

Example

Please give me the *true facts* in the case.

The words *true* and *facts* are saying the same thing. *Facts* are *truths*. Substituting *truths* for *facts* makes the redundancy obvious.

Please give me the *true truths* in the case.

Exercise 7-5 will help you recognize redundancies.

EXERCISE 7-5

All of the following sentences have redundancies. Delete the unneeded words to make the sentences correct.

1. We tried your new idea today. We plan to repeat it again soon.

2. The machine was repaired by 12 p.m. midnight.

3. Management trainees often work in pairs of two.

4. Planning in advance will be helpful.

5. Sales increased up during the summer season.

6. Court testimony should be the honest truth.

7. Our consensus of opinion was positive.

8. Prices were reduced down.

9. Two pipes must be joined together.

10. Sam asked him to return back the equipment.

EXERCISE 7-6

Edit the memo draft that appears on the top of the next page without changing the meaning.

1. Reduce four wordy phrases to one word each.
2. Take out four redundant words and two other unneeded words.
3. Remember to use appropriate revision symbols.
4. Your instructor may ask you to type the edited memo on a separate sheet of paper.

NAME ▬▬▬▬▬▬▬▬▬▬▬▬▬▬▬▬▬▬ DATE ▬▬▬▬▬▬

MEMORANDUM

To: Cathy Eisenberg
From: Hector Sabines
Date: August 18, 198-
Subject: Reducing Down Stationery Cost

Prior to the time that we bought a word processing
system, all letters were typed up on stationery.
In almost all instances, we didn't use verification
drafts because retyping a correct letter again was
more expensive than the stationery used on letters
that had to be redone over. Frequently, new errors
occurred when letters were retyped again.

To reduce down stationery cost, please print out a
verification draft on plain paper should it be that
the same letter is going to five or more people.
Ask the originator to edit the draft. In the event
that changes are needed, they can be made before
expensive stationery is used.

                              HS

fp

Overuse of Words

Another word-use problem is the unnecessary repetition of words or groups of words. This repetition can be in the same sentence or anywhere in the message.

Sometimes repetition errors are not detected by the author because the repeated words are part of the author's usual vocabulary. However, these repetitions "stick out" to the reader. Stick-outs quickly take the reader's attention away from the message; thus, maximum effectiveness is lost.

Being aware of needless repetition errors will help you find them. This awareness plus practice will cause the errors to stick out for you during editing.

Word processing equipment can locate needless repetitions automatically, but the operator must decide how to correct the error. For example, suppose you are editing the first page of a three-page report and the word *very* appears twice. Is this word overused? Find out by doing a global search. (Global functions are briefly discussed in Chapter 2.)

The global search feature can count the occurrences of a specific arrangement of strokes. The arrangement can be words, groups of words, or figures. Suppose the global search revealed ten occurrences of *very*.

Ten *verys* in a three-page report are probably too many. Using the global search feature again, find the occurrences. This command stops the cursor at every *very*. The operator must then decide how to correct the overuse of *very*.

Several methods for correcting needless repetition appear on the next page.

NAME ▬▬▬▬▬▬▬▬▬▬▬▬▬▬▬▬▬▬▬▬▬▬▬▬▬ **DATE** ▬▬▬▬▬

Problem Sentence

   Thank you *very* much for the *very* thoughtful gift.

1. Delete all occurrences. (*Much* must be omitted.)
      Thank you for the thoughtful gift.
2. Delete one occurrence.
      Thank you very much for the thoughtful gift.
3. Use synonyms.
      Thank you very much for the extremely thoughtful gift.
4. Reword the sentence.
      Thank you very much for such a thoughtful gift.

EXERCISE 7-7

In each numbered item, underline the needless repetitions. Next, rewrite the sentences as specified.

1. I am most appreciative of your most generous donation to this most deserving cause.

   a. Delete all occurrences.

   _____

   _____

   _____

   b. Delete two occurrences.

   _____

   _____

   _____

   c. Delete one occurrence. Use a synonym for one occurrence.

   _____

   _____

   _____

2. John thinks that the plan will work. He thinks it will save money.

   a. Reword the sentences. (Hint: Combine the sentences.)

   _____

   _____

3. In my opinion, Steve will do a good job in the new position. Steve, in my opinion, may not want to leave the good job he has.

   a. Select appropriate methods to correct *all* the needless repetitions.

   _____

   _____

   _____

   _____

NAME ■■■■■■■■■■■■■■■■■■■■■■■■■■■■■■■■■■ **DATE** ■■■■■■■■

REVIEW EXERCISE 7-A

Edit the following memo without changing the meaning or removing the goodwill. Delete all the words that aren't needed. Your instructor may ask you to type the edited memo on a separate sheet of paper. You should then be sure that the resulting message has correct grammar and punctuation.

MEMORANDUM

To: Anna Kriparos, Office Manager

From: Robert Handler, Mail Room Supervisor

Date: September 9, 198-

Subject: Mail Room Procedures

Please help us to serve all of you and all of our customers better.

Each afternoon at about 3:50 p.m., the mail room has a problem crisis. All departments bring their mail somewhere close to about 3:45 each afternoon. To meet the post office 4:20 departure deadline, we must leave here by no later than 4:05 each afternoon to have the mail at the post office to meet the 4:20 p.m. deadline at the post office.

Processing so very much mail in such a very little time is impossible and just can't be done. Starting tomorrow, Sam Vickers will go to each office to pick up and get your mail between 2:15 and 3:15. Bring all of the rest of the mail that is done later or after that time to the mail room to us as you usually do.

We feel this daily pick-up of mail every afternoon will help you and us.

RH

dk

REVIEW EXERCISE 7-B

Use the techniques you have learned so far to edit the letter on the next page.

1. *Edit misspellings and mechanical errors using revision symbols.*
2. *Revise paragraph lengths appropriately.*
3. *Change redundancies and repetitions.*

NAME ▬▬▬▬▬▬▬▬▬▬▬▬▬▬▬▬▬▬▬▬ DATE ▬▬▬▬▬

ACCOUNTING SERVICES, INC.

POST OFFICE BOX 1699 • HAMMOND, INDIANA 46325-4423 • (219) 555-9723

May 23, 198-

Mrs. Jolene Baskins
Baskins and Alvarez Company
2389 Dunmore Avenue
Ann Arbor, Mi 48103-2236

Dear Mrs. Baskins:

Teresa J. Laporte asked me to to relate her
work experience and her qualifications for your
office managers position. I recommend Teresa very
highly. Teresa has worked holidays and summers
while attending Fremont University. Last summer,
Teresa replaced vacationing employees. Every two
weeks, she learned a different job. These different
jobs gave Teresa experience in handling many different office tasks. Teresa operated the switch-
board, operated the word processing equipment, and
operated the computer terminal.

Teresa's experience and human relations skill
combined together with her leadership ability and
technical knowledge indicate her office management
potential. Please let me know if you need other
infromation. I'm very eager to help Teresa get
this very challenging position.

Very truly yours,

Samuel P. Goddard
Vice President

rl

CHAPTER

Sentences

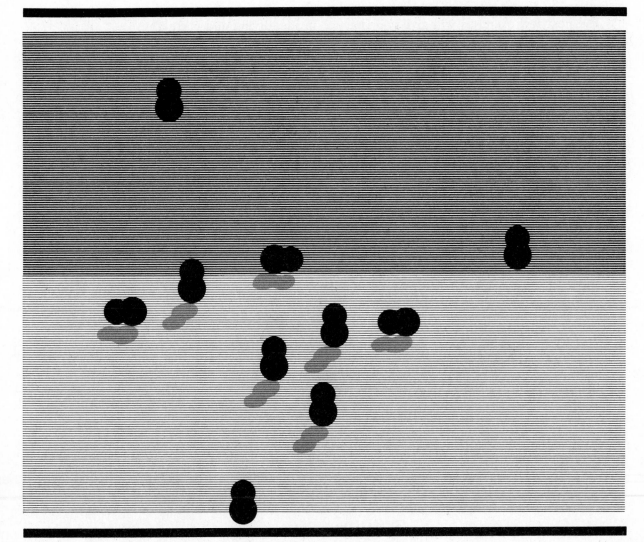

Sentence Fragments

A **sentence** is a group of words that uses a subject and a predicate to express a *complete* thought.

A **fragment** is an incomplete sentence that is incorrectly used as a whole sentence. It does not express a complete thought.

A fragment may be a phrase or a dependent clause. Even though it has a subject and a predicate, a dependent clause is still incomplete.

Phrase

To go to the conference on May 17.

Clause

Because they will meet you at the conference.

Sometimes a fragment is used instead of a sentence to achieve a special effect.

Of all the people to call in late!

Generally, however, business communications should be written in sentences, not fragments. Fragments can be imprecise and confusing.

The next few exercises will give you practice in identifying and correcting fragment errors.

EXERCISE 8-1

Each numbered item below is either a sentence or a fragment. Write S for sentence or F for fragment next to the item. For all sentences, underline the main clause subject once and the main clause predicate (verb only) twice. Write S (for sentence) next to items having both subject and predicate. The first two items are done for you.

_____S_____	1. Joseph talked with the manager.
_____F_____	2. Before the beginning of the month.
_____	3. After July 1.
_____	4. Of the many people attending.
_____	5. One of the many applicants for the position.
_____	6. Mario learned to operate the computer.
_____	7. Who being honest and sincere.
_____	8. If Juanita had asked for the transfer.
_____	9. Before being promoted to assistant manager.
_____	10. Whom we have recently hired.
_____	11. Because the amount of correspondence has increased, we bought a word processor.
_____	12. That something should be done.

NAME ■■■■■■■■■■■■■■■■■■ DATE ■■■■■■■

One way to correct fragment errors is to combine the fragments with a related sentence.

1. Identify the fragment.
2. Find the related sentence.
3. Add the fragment to the sentence at the beginning, in the middle, or at the end.
4. Revise the capitalization.
5. Revise the punctuation.

Exercise 8-2 will help you identify the fragment and find the related sentence.

EXERCISE 8-2

When you are combining dependent and independent clauses to eliminate a fragment, make sure the two clauses are related. Next to the item number, write the letter of the dependent clause that best relates to the independent clause.

Dependent Clauses

a. As soon as he is 65
b. Even though they will use equipment with correcting features
c. Because they are flammable
d. That are needed by 8:30 Friday morning
e. Who have sales experience

Independent Clauses

_____ 1. Some raw materials must be stored separately.

_____ 2. Matt, our engineer, plans to retire.

_____ 3. Typists should be accurate and efficient.

_____ 4. Interviews for the sales manager's position are for qualified applicants.

_____ 5. Shirley Cohen will finish the letter and the report.

After combining a fragment with a sentence, you need to revise the sentence punctuation. In the newly formed sentence, use commas to separate these elements from the rest of the sentence:

Introductory words, phrases, and clauses.
Nonessential words, phrases, and clauses.
Parenthetical words, phrases, and clauses.

Before completing the next two exercises, review the Reference Section on the use of the comma.

EXERCISE 8-3

After reviewing the use of commas with phrases and clauses in the Reference Section, use revision symbols to show the needed commas. If the item does not need commas, put a C next to the item number.

_____ 1. After studying the problem they made their recommendation.

2. The report in your opinion is not complete.

———————— 3. We hired the accountant who had experience.

———————— 4. If the weather does not improve the building will not be finished as scheduled.

———————— 5. Having seen all the available homes they bought the one on Capanella Street.

———————— 6. Because Ricardo likes to travel he requested a transfer to our plant in Germany.

———————— 7. After our plans are final I'll call the branch managers.

———————— 8. I'll write Pete and Jenny after a decision is made.

———————— 9. Even though the matter is sensitive it must be handled.

———————— 10. As mentioned on the phone yesterday was a record day for sales.

EXERCISE 8-4

Correct each fragment error by combining the fragment with the independent clause.

1. Identify the fragment as a block.
2. Move the block to achieve the sentence order specified.
3. Mark capitalization and punctuation changes.
4. Write the resulting sentence.

Note: *For item 4, it is not necessary to block the fragment. The first item is done as an example.*

1. Position the fragment within the clause (after accountant).

After taking several courses. The accountant passed the Certified Public Accountant examination. *move*

The accountant, after taking several courses, passed the Certified Public Accountant examination.

2. Position the fragment before the clause.

Please ship the order today by truck. If the weather improves.

————————————————————————————————

————————————————————————————————

3. Position the fragment within the clause (after *price*).

As advertised on the radio. The price was wrong.

————————————————————————————————

————————————————————————————————

4. Position the fragment before the clause.

For security reasons. Lock the doors immediately after the bank closes.

————————————————————————————————

————————————————————————————————

5. Position the fragment within the clause (after *article*).

   In my opinion. The article seemed biased.

   _____

   _____

6. Position the fragment at the end of the clause.

   After lunch. I'll try to call him again.

   _____

   _____

Run-On Sentences

When you are editing business messages, eliminate run-on sentences. A **run-on sentence** erroneously joins independent clauses. Run-on sentences are formed in two ways.

1. Punctuation between the independent clauses is missing.

   Wrong   Mr. Carson is at lunch he will return at 12:30.

2. Punctuation used between independent clauses (comma) is too weak.

   Wrong   Mr. Carson is at lunch, he will return at 12:30.

   Below are three ways to correct run-on sentences. Choose the one that best suits the message.

1. Use a period to make two sentences.

   Mr. Carson is at lunch. He will return at 12:30.

2a. Use a semicolon between two closely related clauses.

   Mr. Carson is at lunch; he will return at 12:30.

   b. Use a semicolon, a conjunctive adverb, and a comma between the clauses.

   Mr. Carson is at lunch; however, he will return at 12:30.

3. Use a comma and a conjunction between the clauses.

   Mr. Carson is at lunch, and he will return at 12:30.

NAME ▬▬▬▬▬▬▬▬▬▬▬▬▬▬▬▬▬ DATE ▬▬▬▬▬▬

EXERCISE 8-5

Using capitalization, insertions, and punctuation revision symbols, show the specified correction for each run-on sentence. Write the entire corrected item.

1. Use a period.

   a. John is on a sales trip he will be back tomorrow.

   _____

   _____

   b. Order stationery from Logo Company order pens from Taft Office Supply, Inc.

   _____

   _____

   c. She finished college yesterday her raise begins today.

   _____

   _____

2. Use only a semicolon.

   a. The desk was damaged the chair was not.

   _____

   _____

   b. Katrina dictated the letter her secretary typed it.

   _____

   _____

3. Use a semicolon, a conjunctive adverb (*however* or *likewise*), and a comma.

   a. Lyn applied for an accounting position he accepted a management position.

   _____

   _____

   b. Sales are increasing profits are increasing.

   _____

   _____

4. Use a comma and a coordinating conjunction (*and*, *but*, or *or*).

   a. Mr. Dunhill's work was exceptional he got a promotion.

   _____

   _____

   b. More people were hired we are still behind schedule.

   _____

   _____

   c. We can work all day Friday we can work half a day Friday.

   _____

   _____

Too Many Phrases and Clauses

Prepositional phrases and clauses are useful in business writing. However, too many prepositional phrases or too many clauses can interrupt the message rhythm and clarity.

Below are some ways to improve the wording of phrases and clauses.

1. Use a modifier.

   sales manager for the region becomes *regional sales manager*

2. Use a possessive. (Note apostrophe.)

   home of our manager becomes *our manager's home*

3. Use an appositive. An **appositive** is a word or term that refers to the noun or pronoun that precedes it. (Note commas around the appositive.)

   Charles, who is our stockbroker, said becomes *Charles, our stockbroker, said*

4. Use better wording.

   He came on the first day of last week. becomes *He came last Monday.*

EXERCISE 8-6

Without changing the meaning of the sentences, remove as many clauses and prepositional phrases as possible. Use revision symbols to mark changes. Reword as necessary.

1. Please telephone the director of safety in our plant today.

   _____

2. Emilio purchased a typewriter with memory last week.

   _____

3. The meeting is in the office of the manager at 3:15.

   _____

4. Most of the employees in the shipping department agree with us.

   _____

5. The sale starts on the first day of the first month of the calendar year.

   _____

6. Please list the merchandise that was damaged.

   _____
   _____

NAME _____ DATE _____

7. Dr. Raymond Knight, who is our management consultant, recommended procedures that would cut costs.

_____

_____

8. Ed sold eight camping trailers that were new and five cars that were used.

_____

9. We are considering the bid for carpeting that was submitted by Dunn Company, which is located in Cleveland, Ohio.

_____

_____

10. Our accountant suggested that we buy bonds that are long-term.

_____

Sentence Length

You can improve business communications by adjusting sentence length. Using all short sentences has a choppy effect, which bores readers. Using all long sentences makes the message hard to understand. What is the solution?

The solution is variety. Use the following guideline for sentence length. Most sentences should range from ten to twenty words. Obviously, consider the reader when you are deciding on suitable sentence length. The following exercises will clarify this point.

Example 1

As personnel manager, you are inviting children (grades 1 to 6) to summer day camp. Consider your reader. Will most sentences be short? Yes. Even though the sentences are short, should they all be about the same length? No.

EXERCISE 8-7

The numbered sentences on the next page are an excerpt from the personnel manager's letter in Example 1 in the preceding text. Count the words and numbers in each sentence. Put the total in the answer blank provided. Then, complete the graph. Correctly completed, the graph should give you a visual impression of the sentence length of this letter excerpt. Notice that the sentence length varies.

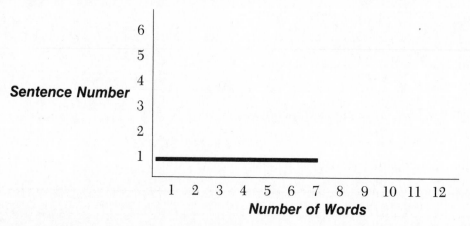

Item 1 is your example.

___7___	1. Are you looking forward to summer vacation?
_____	2. School will be out soon.
_____	3. What will you do with that extra time?
_____	4. We have a suggestion.
_____	5. You have a special invitation to attend our day camp program.
_____	6. This program is for children of our employees.

Example 2

If you are writing a letter to the company president, will most sentences be longer than those in Example 1? Yes. Even though the sentences are longer, should they all be about the same length? No.

EXERCISE 8-8

The sentences below are an excerpt from the letter to the company president in Example 2 in the preceding text. Count the words and numbers in each numbered sentence. Put the total in the answer blank provided. Then, complete the graph. Correctly completed, the graph should give you a visual impression of the sentence length of this letter excerpt. Does the sentence length vary?

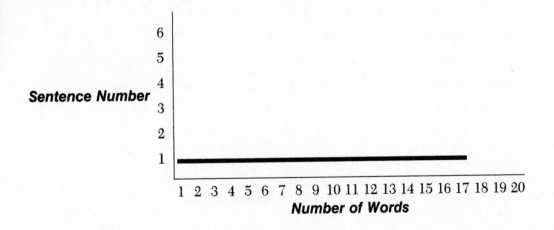

Sentence Number

6
5
4
3
2
1

1 2 3 4 5 6 7 8 9 10 11 12 13 14 15 16 17 18 19 20
Number of Words

Item 1 is your example.

___17___	1. As you requested in our November 5 meeting, I have examined the cost budget in all departments.
_____	2. The objective of this study was to make recommendations for reducing costs.
_____	3. My complete report, enclosed with this letter, summarizes the major concerns needing immediate attention.
_____	4. The main problem is the quality of raw materials.
_____	5. Some defects are evident soon.
_____	6. Other defects do not show up until final processing which adds to the expense.

NAME ━━━━━━━━━━━━━━━━━━━━━━━━━━━━━━━ DATE ━━━━━━━

EXERCISE 8-9

Complete Parts A, B, C, and D.

Part A

Count the words, numbers, and abbreviations in each sentence and put the total in the answer blank provided.

———————————   1. You are encouraged to attend a seminar.

———————————   2. The seminar deals with our fringe benefits.

———————————   3. The date will be Monday, March 8.

———————————   4. It will be in the conference room.

———————————   5. The seminar lasts from 8:30 to 10:30 a.m.

———————————   6. The seminar will be repeated at 1:30 p.m.

———————————   7. You may attend either one of the seminars.

Part B

Answer Yes *or* No *to the following questions.*

———————————   1. Do the sentences have enough variety in length?

———————————   2. Are most of the sentences too long?

———————————   3. Are most of the sentences too short?

———————————   4. Should this message be edited for appropriate sentence length?

Part C

Edit sentences 1 to 7 from Part A of this exercise for variety in length. Combine the seven sentences into three sentences. In editing for sentence length, you will probably make other improvements in wording. Write your sentences below.

1. _____

   _____

   _____

2. _____

   _____

   _____

3. _____

   _____

   _____

NAME ██   DATE ████████████

Part D

Complete the graph for the sentences in Part C. Then answer the following questions.

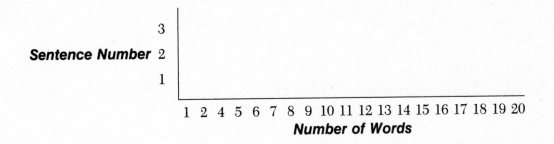

Sentence Number

3

2

1

1 2 4 5 6 7 8 9 10 11 12 13 14 15 16 17 18 19 20
Number of Words

1. Are your sentences varied in length? _____

2. In what other ways did you improve the sentences?

_____

_____

_____

_____

_____

_____

_____

_____

Superfluous Paragraphs

While editing for conciseness, sometimes you will find an entire paragraph that is useless. This useless paragraph can detract from the intended message or cause other problems.

To delete a useless paragraph, identify it as a block and use the delete symbol.

Example

This is a useless paragraph. Delete it. Deleting a useless paragraph will improve your message.

Exercise 8-10 will help you to become aware of superfluous paragraphs.

EXERCISE 8-10

Read the memorandum that appears on the following page and complete the items that come after it.

NAME ━━━━━━━━━━━━━━━━━━━━━ **DATE** ━━━━━

MEMORANDUM

```
To:       Camille Finkelstein
From:     Bill Werner
Date:     May 18, 198-
Subject:  Vacation Request
```

Your vacation request for July 24-31 has been approved. After completing that big marketing project, you certainly deserve a break.

You were lucky! People with more seniority didn't apply for that week. Tom Moyer asked for July 1-7; Julie Brown asked for July 8-15; and Will Baxter asked for July 16-23.

Best wishes for a restful vacation.

```
                              BW

rd
```

1. Which *one* sentence expresses the reason for writing the memo?
   Sentence _____ in paragraph _____

2. Which two sentences are for goodwill purposes?
   Sentence _____ in paragraph _____ and sentence _____ in paragraph _____

3. Which paragraph can be omitted without the loss of message clarity and goodwill?
   Paragraph _____

4. Identify the useless paragraph as a block and delete it.

5. Superfluous paragraphs can cause problems. What problem is created by the superfluous paragraph in this memo?

NAME ▬▬▬▬▬▬▬▬▬▬▬▬▬▬▬▬▬▬▬▬▬▬▬ **DATE** ▬▬▬▬

REVIEW EXERCISE 8-A

Edit the vacation notice below for variety in sentence length and for sentence errors. Rewording will be necessary. Your instructor may ask you to type the revised notice on a separate sheet of paper.

V A C A T I O N   N O T I C E

KIRBY MANUFACTURING PLANT

AT THE END OF THE SECOND SHIFT ON THE FIRST DAY OF JULY, WHICH IS ON A FRIDAY, KIRBY MANUFACTURING PLANT WILL CLOSE FOR VACATION, OPERATIONS WILL RESUME WITH THE FIRST SHIFT ON MONDAY, JULY 11.

SUPERVISORS WILL DISTRIBUTE VACATION BONUS CHECKS DURING YOUR LAST SHIFT ON FRIDAY. YOU WILL RECEIVE 5 PERCENT OF YOUR SALARY FOR THE FIRST FIVE MONTHS OF THIS YEAR. PLUS $20 FOR EACH YEAR WITH THE COMPANY.

BEST WISHES FOR A SAFE AND RESTFUL VACATION!

J. B. KAHN
MANAGER

NAME ━━━━━━━━━━━━━━━━━━ DATE ━━━━━━

REVIEW EXERCISE 8-B

Use revision symbols to edit the social-business letter below. Your instructor may ask you to type the edited letter on a separate sheet of paper.

819 Wellman Lane
New Castle, PA 16101
September 2, 198-

Harrison's Merchandise, Inc.
3721 Newell Avenue
New Castle, PA 16102

Ladies and Gentlemen:

Please ship out to me the following items by Speedy Package Service for arrival by September 10th: one No. 6130 Bentley Pocket Watch in white gold ($75), one No. 6131 Bentley Watch Chain in white gold ($25), one No. 7130 Bentley Pocket Watch in yellow gold ($75), and one No. 7131 Bentley Watch Chain in yellow gold ($25). The total is $300.

September 15 is my twin brothers' twenty-first birthday. I wanted to get them something special. Robert likes white gold, and Steven likes yellow gold.

Charge the total plus tax and shiping expense to my Harrison's account. Thank you.

Sincerely,

Alex Simpson

CHAPTER

Consistency of Fact, Treatment, and Usage

Consistency of Fact

When you are editing, always look for facts that disagree. Here are some ways facts can be contradictory.

1. Contradiction between the message and an established fact that doesn't appear in the message.
2. Contradiction within a phrase or a sentence.
3. Contradiction within a message but not within the same sentence.
4. Contradiction between the message and another message such as an enclosure.

EXERCISE 9-1

Use consecutive letters (A, B, C, and so on) to block and query the part of each item that contradicts an established fact. Write the established fact in the blank provided. If there is no contradiction, write OK in the blank. The first item has been completed as an example.

1. letter dated April 31 [X] ?

   April has 30 days.

2. 12:64 p.m.

   _____

3. September 31

   _____

4. Invoice dated 21/10/84

   _____

5. 9:30 a.m.

   _____

6. Corvettes cost $1,120.

   _____

One way to clarify a contradiction is to check a reliable source. Suppose the following sentence appeared in a message.

The cost increased from $300 to $275.

You would not know if *increased* should be changed to *decreased* or if $275 should be changed to some other figure—perhaps, $375. You would only know that a change was needed, and you would query the contradiction.

The cost |increased| from |$300 to $275|. A ? B

Next, suppose you consulted a reliable source, such as another document or a knowledgeable person, and determined that the cost had actually decreased. You would make the appropriate change and then cross out the query.

The cost ~~increased~~ *decreased* from |$300 to $275|. A ? B

Exercise 9-2 provides practice in querying a contradiction, consulting a reliable source, and resolving the query.

NAME ━━━━━━━━━━━━━━━━━━━━━━━━━━━━━━━━━━━ DATE ━━━━━━━━

EXERCISE 9-2

For each item that has a contradiction: Use consecutive letters (A, B, C, and so on) to block and query the contradiction. Check a reliable source. Use the list provided. Use revision symbols to mark the needed change. Remember to mark any mechanical changes resulting from the revisions. Write the revised item.

The first item has been completed as an example.

Reliable Source

The drapes were shortened. The price decreased.
It comes in two colors. The sales staff increased.
Joe is not a manager.

1. price ~~increased~~ *decreased* from $20 to $15 ⊠ ? ⊠
   price decreased from $20 to $15

2. two managers, Abe, Joe, and Sue

3. lengthened the office drapes
   from 84 to 81 inches

4. It comes in three colors: red and blue

5. The sales staff decreased from three to five.

6. The speed limit decreased
   from 60 mph to 55 mph.

EXERCISE 9-3

The letter draft on page 116 has contradictions that are not in the same sentence.

Part A

Using the marked blocks in the letter, complete items 1 to 4.

1. Which two blocks contradict block L? ☐ ☐ ? L

2. Which two blocks agree but contradict block D? ☐ ☐ ? D

3. Which two blocks, when added together, disagree with block I? ☐ ☐ ? I

4. These three blocks should be the same, but all three are different. ☐ ? ☐ ? ☐

Part B

Complete items 5 and 6.

5. Checking with a reliable source reveals that these blocks are correct: blocks A, B, E, F, G, H, and J. Block D should be *48*. Use revision symbols to mark all needed changes.

6. Your instructor may ask you to type the revised letter on a separate sheet of paper.

NAME　　　　　　　　　　**DATE**

Sherman Printing Services

2200 North Saguaro • Mesa, Arizona 85201

(602) 555-8157

|A
|April 12, | 198-

Mrs. Wanda Tallchief
Mesa Village Mall, Inc.
Post Office Box 3482
Mesa, AZ 85205

Dear Mrs. Tallchief:

This letter confirms our phone quotation for
printing your |50-page| policy manual. The |$12,000|
total includes the artwork on the cover and on
page |58.| Itemized costs will be as follows:

  Keyboarding |(50 |pages at $2) $ 100
  Editing (10 hours at $10)   100
  Copying (500 copies at $2) |1,000
    Total |$1,200|

You will get an approval draft |two| weeks after
you give us the original, and your manuals will be
delivered |one| week after draft approval. Assuming
we get immediate draft approval, this schedule allows
only |four| weeks from original to delivered manuals,
which will help meet your deadline.

Our quotation is effective for only |30 days|
because of an expected price increase in paper. To
assure this |$2,100| price, please order by |June 12.|

    Sincerely,

    R. David Sherman

jb

EXERCISE 9-4

Complete Parts A, B, and C.

APRIL						
S	M	T	W	T	F	S
		1	2	3	4	5	6
7	8	9	10	11	12	13
14	15	16	17	18	19	20
21	22	23	24	25	26	27
28	29	30				

Part A

Block and query the contradictions between the memo draft and the calendar and between the memo draft and the report. Use the calendar shown here. Put queries in right margin of the memo.

MEMORANDUM

To: Marsha Dillingham
From: Jean Larson
Date: March 25, 198-
Subject: Sales Report

Enclosed is the report for the first four months of the calendar year. Our operations were within the budget for the first two months and over the budget for March, giving us a net operating loss of $4,000.

At our meeting Monday, April 5, we will discuss ways to reduce operating costs. The meeting is at 10 a.m. in my office.

                    JL

cb
Enclosure

BUDGET REPORT
January, February, and March, 198-

Month	Budgeted Amount $	Actual Expense $	(Over) Under $
January	50,000	49,000	1,000
February	40,000	38,000	2,000
March	60,000	66,000	(6,000)
Net Operating Loss			(3,000)

Part B

Assuming that the calendar and the report are accurate and that the meeting is the first Monday in April, use revision symbols to mark needed changes in the memo. Cross out each query after the revision is marked.

Part C

Your instructor may ask you to type the revised memo on a separate sheet of paper.

NAME ━━━━━━━━━━━━━━━━━━━━━━━━━━   DATE ━━━━━━

Consistency of Treatment

Similar items should be treated the same way. For example, if a word is spelled out in one sentence, it should not be abbreviated in another part of the message. Finding and correcting items treated inconsistently is part of editing. The Reference Section shows some of the ways inconsistencies appear in business writing. Studying this section will help you with the next few exercises.

EXERCISE 9-5

Use revision symbols to make the one *inconsistently treated part consistent with the treatment of other parts of the item.*

1. Dr. Hyatt's books—<u>Economic History</u>, ECONOMIC FORECASTS, and ECONOMIC TRENDS —are selling well.

2. Frank Garcia, Joe Lightfoot, and Ms. Chandra Gupta are on the program.

3. While I'm away, these are your responsibilities:

   a. Answer routine letters.

   b. Schedule part-time workers.

   C. Prepare the usual reports.

4. Outline:

   ### TIME MANAGEMENT

      I. List the tasks that must be done today.

     II. List the tasks that must be done tomorrow.

     3. List the tasks that must be done this week.

5. Arrange appointments with secretarial applicants with these qualifications:

   1. Her minimum typing speed—60 words a minute.

   2. Minimum shorthand speed—80 words a minute.

   3. Minimum on aptitude test—80 percent.

6.

<u>Letters</u>	Memorandums	<u>Reports</u>
5	9	2
4	8	3
2	5	4

7. Each applicant must be a college graduate with a business related major. They must have good communication skills.

8. Responsibilities include answering the phone, greeting our visitors, and type purchase orders.

9. Cheryl, Dave Rodriquez, and Janet were hired.

10. My plans are as follows:

    1. Graduate from college

    (2) Get a job that encourages advancement.

    (3) Attend graduate school at night.

EXERCISE 9-6

In the memo draft, find five sets of items that were treated inconsistently. Next, resolve the inconsistency by choosing either treatment. See the example below. Use revision symbols to mark for consistent treatment. Be sure to make any other corrections needed as a result of editing the inconsistencies.

Steve Arden and Ms. Linda Fanning will be promoted.

Mr. ⌃ Steve Arden and Ms. Linda Fanning will be promoted.

MEMORANDUM

To: Brandon Martin
From: J. B. Batacan
Date: March 7, 198-
Subject: Mallory Autograph Session

Ms. Stella R. Mallory will be here April 8 from 2:30 to 5:30 p.m. to autograph copies of her new book, <u>Success at Home</u>. To prepare for her visit, please do the following:

First, order 500 copies of Mrs. Mallory's new book and 100 copies of her other book, SUCCESS AT WORK.

2. Draft a press release announcing her visit.

Fred is meeting her flight (Sky Airlines No. 837) at 12:45 p.m. and to drive her back to the airport. Her flight, Sky Airlines No. 842, leaves at 6:45 p.m.

JBB

rm

Consistency of Usage

Consistency of usage involves agreement between parts of speech.

1. Subjects and verbs must agree in number and person.
2. Pronouns and antecedents must agree in gender, number, and person.

The next exercise will give you practice in editing for consistency in usage. For a review of subject/verb agreement and pronoun/antecedent agreement, see the Reference Section.

NAME ▬▬▬▬▬▬▬▬▬▬▬▬▬▬▬▬▬▬▬▬ **DATE** ▬▬▬▬▬

EXERCISE 9-7

Change verbs or pronouns to make usage consistent. If usage is consistent, write OK beside the item number. Items may have more than one inconsistency. Remember to use revision symbols to mark the changes.

_____ 1. New machines was ordered by the purchasing agent.

_____ 2. The accountant, the sales representative, and the client disagreed on credit terms. Both of them have an interest in the decision.

_____ 3. Most stockbrokers, in my opinion, wants to make a profit.

_____ 4. Dean or Glen have the camera.

_____ 5. Mr. Kerns and Mr. Hill were early.

_____ 6. The manager or her assistants is checking on the matter.

_____ 7. Peter nominated herself for president.

_____ 8. Jan and Steve apologized. They said that their actions were inappropriate.

_____ 9. Cathy and Tom is being considered for the position.

_____ 10. Neither Jill nor Ann want their individual responsibilities changed.

REVIEW EXERCISE 9-A

Six bids (three for painting and three for carpeting) that were submitted to Harrison Realty, Inc. follow. The bids (assume that they are correct) were used as source documents for the memo on page 121. Use revision symbols to edit the memo. There is no need to block and query.

CARPET BID

My bid for the carpeting contract as specified on Carpet Bid Sheet 820 is $3,750

Signature *Cecil Culp*
Cecil Culp
Date April 9, 19--
Company Portland Floor Covering
Address Post Office Box 68089
Portland, OR 97266

PAINT BID

My bid for the painting contract as specified on Paint Bid Sheet 810 is $5,850

Signature *Charles B. Fuller*
Charles B. Fuller
Date April 8, 19--
Company The Paint Center
Address Post Office Box 25842
Portland, OR 97225

CARPET BID

My bid for the carpeting contract as specified on Carpet Bid Sheet 820 is $3,500

Signature *Kathryn Kubas*
Kathryn Kubas
Date April 8, 19--
Company Luxury Carpets, Inc.
Address Post Office Box 42887
Portland, OR 97242

PAINT BID

My bid for the painting contract as specified on Paint Bid Sheet 810 is $5,500

Signature *J. B. Jordan*
J. B. Jordan
Date April 7, 19--
Company Jordan's Paint Factory
Address Post Office Box 10999
Portland, OR 97210

NAME _____   DATE _____

CARPET BID

My bid for the carpeting contract as specified on Carpet
Bid Sheet 820 is ___$3,650___

Signature _*Joseph Gaviola*_
Joseph Gaviola

Date ___April 7, 19--___
Company ___Carpet Sales & Service___
Address ___Post Office Box 13925___
___Portland, OR 97213___

PAINT BID

My bid for the painting contract as specified on Paint
Bid Sheet 810 is ___$5,700___

Signature _*Mary C. Cross*_
Mary C. Cross

Date ___April 10, 19--___
Company ___Portland Paint & Glass___
Address ___Post Office Box 14521___
___Portland, OR 97214___

MEMO

To: Carmen Antonio
From: Hannah Vannoy
Date: April 14, 198-
Subject: Carpet and Paint Bids

Enclosed is the bids for carpet (Bid Sheet 820)
and painting, Bid Sheet 810, our ten duplex
apartments.   The bids are listed below:

CARPET BIDS

1.   Carpet Sales & Service $3,650
2.   Luxury Carpets, Inc. $3,500
3.   Portland Floor Covering $3,750

Paint Bids

a.   Jordan's Paint Factory $5,500
b.   Portland Paint & Glass $5,700
c.   The Paint Center $5,750

Luxury Carpets, Inc., had the lowest carpet bid,
$3,500; and The Paint Center had the lowest paint
bid, $5,500--a total of $10,000.

Please call the companies with the lowest bids.
Ask it to begin within two weeks but no later than
the first of April.

HV

k1
Enclosure
cc: Edward Jones and Ms. Erin Parsons

NAME ▬▬▬▬▬▬▬▬▬▬▬▬▬▬▬▬   DATE ▬▬▬▬

REVIEW EXERCISE 9-B

Sometimes business writers dictate essentially the same letter again and again because the same kind of response is needed often. Form letters can save time. After the form letters are developed, the business person gives address information, form letter number, and any special instructions to the keyboarder.

Part A

The drafts of two form letters follow. Use revision symbols to edit both letters for mechanical errors.

Form Letter 1:   Welcome--New Credit Customer

_____
                     Date

_____
_____
_____
_____

Dear _____;

     Thank you for using your new credit account.
Having the account certainly makes shoping more
convenient.

     Meeting your furniture needs through quality
merchandise and service is our goal. You can choose
from the wide variety of furniture in our showroom,
or you can order that special item that you haven't
been able to fine.

     Let us know when we can serve you again.

                     Sincerely,

          _____ Writer's Name
          _____ Title

     xx

Form Letter 2: <u>Last Payment</u>

_____
                      Date

_____
_____
_____
_____

Dear _____:

     Today, you made the last payment on your account. Thank you for you prompt payments each month.

     We sincerely appreciate you as a customer. To show our our appreciation, we have enclosed a $20 coupon, which are good on your next credit purchase of $200 or more.

     Let us know when you need more furniture. Doing business with you is a pleasure. Shop with us again soon!

     Sincerely,

_____ Writer's Name
_____ Title

xx

Part B

Mr. John K. Williams, manager of Rotella's Furniture Showroom, selected form letters based on the note on page 124. Assuming that the note is correct, edit the form letter requests on page 124.

NAME ━━━━━━━━━━━━━━━━━━━━━━━━━━━━ **DATE** ━━━━━━━━

NOTE

Mr. Williams,

Two customers used new credit accounts today:

       Mr. and Mrs. Jim Torres
       4532 Carmel Drive
       Great Falls, MT 59405

       Mrs. Evelyn D. Stratton
       Gatlin Apartment 632
       605 Parker Drive
       Great Falls, MT 59401

One customer made his last payment today:

       Mr. Martin Massey
       1806 Valleyview Drive
       Great Falls, MT 59404

       JBS
       5/5/8-

REQUEST FOR FORM LETTER

Letter _1_

John Williams _____ May 5, 198-
Requested By   Date

TO

Mr. and Mrs. Jim Torres

4532 Carmel Drive

Great Falls, MT 59405

_____

SPECIAL INSTRUCTIONS Insert as ¶ 3:

You can order that corner hutch

you have been wanting. Enclosed

is the brochure on hutches you

asked us to get for you.

_____   _____
Typed By   Date

REQUEST FOR FORM LETTER

Letter _2_

John Williams _____ May 5, 198-
Requested By   Date

TO

Mrs. Evelyn D. Stratton

Gatlin Apartment 623

605 Parker Drive

Great Falls, MT 59401

SPECIAL INSTRUCTIONS

Salutation —

Dear Evelyn:

_____   _____
Typed By   Date

REQUEST FOR FORM LETTER

Letter _2_

John Williams _____ May 5, 198-
Requested By   Date

TO

Mr. Martin Massey

1806 Valley Drive

Great Falls, MT 59404

SPECIAL INSTRUCTIONS

Use Salutation —

Dear Marty:

_____   _____
Typed By   Date

Part C

Your instructor may ask you to type the revised form letters on separate sheets of paper following any special instructions. John K. Williams is the writer; his title is manager. After typing each letter, sign and date (5/5/8–) the form letter request.

NAME ▬▬▬▬▬▬▬▬▬▬▬▬▬▬▬▬ DATE ▬▬▬▬▬▬

CHAPTER

Sequence of Information

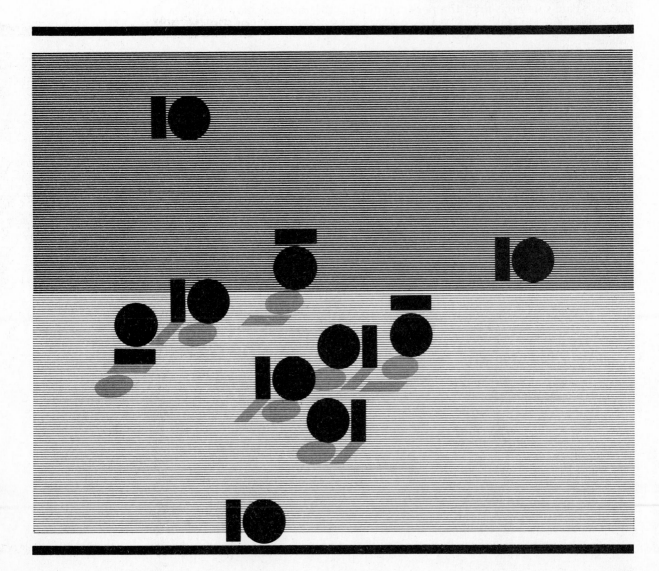

Logical Sequence

Correct sequencing makes business messages easier to understand. Reading poorly sequenced messages is like entering a movie in the middle. It is hard to fit all the pieces together. Usually, putting the information into the proper order is a matter of applying logic.

Proper sequencing—alphabetical, chronological, and numerical order, for example—helps the reader get a clear understanding from the beginning. When order is ignored, there may be problems. Imagine the confusion and frustration caused by the following:

A dictionary *not* in alphabetical order.
A calendar *not* in chronological order.
Book pages *not* in numerical order.

Though not quite as obvious as the above examples, proper sequencing in business messages is important. Alphabetical order can keep the writer from *conveying* bias. Suppose the following sentence appeared in a memorandum.

Paul, Joel, and Amy are being considered for the manager's job.

Because Paul's name is first, the person receiving the message may feel the writer favors Paul for the job or that Amy's name was mentioned last because she is female.

Probably, the writer had no reason for the order; however, the writer did *convey* bias to the readers. Listing names in alphabetical order can solve the problem.

The next few exercises will increase your awareness of proper sequencing in business messages.

EXERCISE 10-1

Show chronological order by numbering each part.

_____	1. a. Night		_____	5. a. August 10, 1984
_____	b. Noon		_____	b. August 8, 1984
_____	c. Morning		_____	c. August 12, 1983
			_____	d. August 20, 1982
_____	2. a. 8:30 a.m.		_____	e. August 25, 1985
_____	b. 7:45 a.m.			
_____	c. 6:30 a.m.		_____	6. a. Monday, April 4
			_____	b. Friday, April 1
_____	3. a. April 1984		_____	c. Thursday, April 7
_____	b. May 1985		_____	d. Tuesday, April 12
_____	c. June 1983		_____	e. Wednesday, April 6
_____	4. a. 1:30 p.m.		_____	7. a. Fourth quarter of the year
_____	b. 12 noon		_____	b. First quarter of the year
_____	c. 11:30 a.m.		_____	c. Third quarter of the year
_____	d. 5:30 a.m.		_____	d. Second quarter of the year

NAME ━━━━━━━━━━━━━━━━━━━━━━━━━━━━━━━━━━━━━ DATE ━━━━━━━━

EXERCISE 10-2

Sequence the following items in numerical order from lowest to highest.

_____	1. a. Invoice 2103
_____	b. Invoice 2105
_____	c. Invoice 2104
_____	2. a. Purchase Order 1613
_____	b. Purchase Order 1610
_____	c. Purchase Order 1612
_____	d. Purchase Order 1611
_____	e. Purchase Order 1614
_____	3. a. Check 305
_____	b. Check 304
_____	c. Check 303
_____	4. a. Invoices due in 5 days
_____	b. Invoices due in 16 days
_____	c. Invoices due in 7 days
_____	d. Invoices due in 30 days
_____	e. Invoices due in 14 days

_____	5. a. Account 403
_____	b. Account 313
_____	c. Account 219
_____	d. Account 519
_____	e. Account 103
_____	6. a. $1,000
_____	b. $10,000
_____	c. $1,500
_____	d. $15,000
_____	7. a. 55.1
_____	b. 52.625
_____	c. 51.787
_____	8. a. $365.18
_____	b. $125.50
_____	c. $142.80

EXERCISE 10-3

Sequence the following items in alphabetical order by numbering each part.

_____	1. a. Abercrombie
_____	b. Deloach
_____	c. Turner
_____	d. Jones
_____	e. Walker
_____	2. a. Frances
_____	b. Ted
_____	c. Joyce
_____	3. a. Smith, Elaine
_____	b. Rosen, Phyllis
_____	c. Bertinelli, Sam
_____	4. a. Seattle
_____	b. Atlanta
_____	c. Dallas
_____	d. Chicago
_____	e. Kansas City

_____	5. a. Mark Franklin
_____	b. Lena Bergon
_____	c. Nancy Harriman
_____	6. a. Word Processing Department
_____	b. Accounting Department
_____	c. Quality Control Department
_____	d. Manufacturing Department
_____	7. a. Eastgate Mall
_____	b. Westgate Mall
_____	c. Silas Creek Mall
_____	d. Suburban Mall
_____	8. a. North Carolina
_____	b. Georgia
_____	c. Florida
_____	d. Alabama
_____	e. South Carolina

NAME ━━━━━━━━━━━━━━━━━━━━━━━━━━ DATE ━━━━━━━━━

EXERCISE 10-4

Sequence the events in order of usual occurrence by numbering each part.

_____ 1. a. I received a reply to my letter.

_____   b. I wrote a letter.

_____ 2. a. Rosa received a typewriter.

_____   b. Rosa ordered a typewriter.

_____   c. Rosa typed a letter.

_____ 3. a. Louis arrived at work at 8:30 a.m.

_____   b. Louis went to lunch.

_____   c. Louis took a break from 10:15 to 10:30 a.m.

_____   d. Louis returned from lunch.

_____   e. Louis left work at 4:30 p.m.

_____ 4. a. The secretary transcribed the letter.

_____   b. The executive dictated the letter.

_____   c. The executive signed the letter.

_____ 5. a. Office productivity increased.

_____   b. The company bought a word processor.

_____ 6. a. Harry broke an appliance during the warranty period.

_____   b. Harry bought an appliance.

_____   c. Harry returned the appliance to the dealer for repair.

_____ 7. a. Acme Video sold the last five televisions (Model 381).

_____   b. Acme Video advertised the five televisions (Model 381).

_____   c. Acme Video had only five televisions (Model 381) left.

_____ 8. a. Mr. Lowell printed the document.

_____   b. Mr. Lowell keyboarded the document into the word processor.

_____   c. Mr. Lowell edited the document on the screen.

_____ 9. a. Ms. Carlucci received a credit card for Haskell's Jewelry Store.

_____   b. Ms. Carlucci applied for credit with Haskell's Jewelry Store.

_____ 10. a. Singh Pharmacy received an invoice.

_____   b. Singh Pharmacy bought merchandise on credit.

_____   c. Singh Pharmacy mailed a check paying for the merchandise.

_____   d. Singh Pharmacy wrote a check to pay the invoice.

Special Sequences

Sometimes accepted style or specific instructions create the need for a special sequence. Here are two examples. A résumé usually lists educational accomplishments and work experience in reverse chronological order.

The United States Postal Service issues specific instructions for the sequence of addresses for maximum efficiency in mail delivery.

    Name
    Company name (or other information)
    Delivery address
    City, State ZIP + 4 code

NAME ▬▬▬▬▬▬▬▬▬▬▬▬▬▬▬▬▬▬▬▬ DATE ▬▬▬▬▬▬

If two addresses are used (a street address and a post office box number), the mail will be delivered to the address on the next-to-the-last line. Order *does* make a difference!

Note

1. Mail with a ZIP code can be machine-sorted to the correct post office. Mail with a ZIP + 4 code can be machine-sorted to the specific carrier who will deliver the item.
2. In the case of two addresses, the street address and the post office box number may have different ZIP codes or Z + 4 codes.

EXERCISE 10-5

Show the specified order by numbering each part. Two examples are given below.

Examples: Reverse order

3	a. 10	2	a. February
1	b. 20	1	b. March
2	c. 15	3	c. January

Reverse order (highest to lowest)

_____ 1. a. $10,000
_____   b. $12,000
_____   c. $20,000

_____ 2. a. 250
_____   b. 600
_____   c. 150

_____ 3. a. Three years
_____   b. One year
_____   c. Two years

Reverse order (most recent first)

_____ 4. a. 1981
_____   b. 1983
_____   c. 1982

_____ 5. a. 1985–86
_____   b. 1981–82
_____   c. 1983–84

_____ 6. a. 10/20/84
_____   b. 10/18/84
_____   c. 10/23/84

Mail delivery to a post office box (Assume that the ZIP + 4 codes are correct.)

_____	7. a. Fitzpatrick Realty Corp.
_____	b. 2285 Aberdeen Way
_____	c. Post Office Box 8210
_____	d. Cheyenne, WY 82001-2187
_____	8. a. Mr. Tim Baxter, President
_____	b. Baxter Manufacturing, Inc.
_____	c. Post Office Box 1821
_____	d. 1864 East First Street
_____	e. Cheyenne, WY 82001-1132

Mail delivery to a street address (Assume that the ZIP + 4 codes are correct.)

_____	9. a. Mr. Rameer Jubran
_____	b. Metropolitan Chemical Corp.
_____	c. 8615 Ranlo Boulevard
_____	d. Post Office Box 1231
_____	e. Cheyenne, WY 82001-3412
_____	10. a. Ms. Alva Bruch
_____	b. Cansler Wholesale Company, Inc.
_____	c. Post Office Box 2516
_____	d. 1908 West Washington Street
_____	e. Cheyenne, WY 82001-9836

EXERCISE 10-6

Use revision symbols to correct errors in the specified sequence. Be sure to mark needed changes resulting from the corrections. If the item is correctly sequenced, write OK _by the item number._

_____   1. Alphabetical order (after _To:_)

To:	Stan Medlin and Patricia Everly
From:	J. B. Gilbert
Date:	April 1, 1984
Subject:	New Employees

_____   2. Alphabetical order

cc: Mr. Brown, Mr. Gibson, and Mr. Applebaum

NAME ▬▬▬▬▬▬▬▬▬▬▬▬▬▬▬▬▬▬▬▬▬▬▬▬▬▬▬▬▬▬▬ DATE ▬▬▬▬▬▬▬

_____ 3. Alphabetical order (by authors' names)

Tanner, Alfred J., *Business Sense*, Patrelli Publishing Company, Inc., New York, 1981.

Brennan, C. Clifford, *Investing Made Easy*, Arnold Publishing Company, Inc., Chicago, 1982.

_____ 4. Chronological order

Fred was promoted to manager in 1983.

Fred began work in 1975.

Fred was promoted to assistant manager in 1980.

_____ 5. Numerical order

Purchase Order 2801 was for equipment.

Purchase Order 2802 was for materials.

Purchase Order 2803 was for supplies.

_____ 6. Reverse order (most recent first for educational background on a résumé)

Central High School, Diploma

University of Kentucky, Bachelor of Science

City Community College, Associate of Arts

_____ 7. Reverse order (most recent first for work experience on a résumé)

1984–85, Sales Clerk, Hoyle's Department Store

1982–83, Cook, Quick-Burger Shop

1980–81, Maintenance Helper, Franco Manufacturing

1985–86, Office Assistant, Metro Life Insurance Company

EXERCISE 10-7

Paragraphs and entire messages should be properly sequenced. In each of the following items, indicate the proper sequence of paragraphs by numbering them consecutively in the blank on the right. Do not change anything within the paragraphs.

Paragraphs

1. Memo

Walter Rosenthal, an employee for ten years, was recommended by his supervisor. He has been here longer than Carmen or Della and has general management experience.   _____

Della Clark, who works in our cost department, applied for the office manager's position. She attended night school for three years and received an office management degree last month.   _____

Carmen Ray, who transferred from our Chicago office, has experience as an office manager but lacks the formal educational background that Della has. She wants to improve herself by taking college courses.   _____

2. Memo

   In 1980, we reported our first profit to our stockholders. Although the profit was small, it began a trend. _____

   In 1982, all machinery was put into peak operating condition. This resulted in less waste, a better quality product, and an additional 2 percent increase in profit. _____

   In 1981, our profit increased by 5 percent. Better efficiency and full cooperation from our personnel lowered operating costs. _____

3. Memo

   Three sales representatives from our office had the highest quarterly sales for our division. Listed according to sales, they are Jim Jacobs, Lisa Manning, and Judy Kelsey. _____

   Jim had the top sales figure of $385,000. He said that satisfied customers are his best sales tool. About a third of his sales came from friends of customers. _____

   Judy's sales were $377,000. This is a 25 percent increase over her last quarter total. She feels the increase is the result of using the sales techniques learned at our seminar. _____

   Lisa's sales totaled $378,000. This figure is impressive; however, knowing that she joined us only six months ago makes it *very* impressive.

Entire Messages

4. Series of Collection Messages
   Messages A, B, and C are sent to customers with overdue bills. In what order should they be sent?

   a. This message has a strong reminder and lists the possible consequences of not paying the overdue invoice. _____

   b. This message reminds the customer of an overdue invoice by suggesting that the invoice may have been overlooked or that the check may have been mailed within the last few days. _____

   c. The message firmly reminds the customer of the overdue invoice, mentions the valuable role credit plays in operating a business, and uses a psychological appeal related to customer fairness and responsibility. _____

5. Series of Sales Messages
   Messages A, B, and C are to be used for the same list of potential customers. The product is expensive, which justifies three messages. After each mailing response, the names of purchasers are removed from the list. In what order should the three messages be mailed?

   a. *Purpose:* attract customer attention, create interest in the product, and sell the product. *Includes:* product information, sales message, and opportunity to buy. _____

   b. *Purpose:* develop interest in product, create desire for product, and sell the product. *Includes:* product information; favorable comparison between product and comparable ones; psychological appeals to ego, intelligence, and pride; sales message; and opportunity to buy. _____

   c. *Purpose:* develop interest in product, develop desire for product, sell the product, and stimulate customer action. *Includes:* product information, inducements for purchasing by certain date, ten-day trial period, a reminder that quantity is limited at this price, easy payment terms, sales message, and opportunity to buy. _____

REVIEW EXERCISE 10-A

Edit the memo drafts below and on the next page for sequence. Use revision symbols to mark changes.

MEMORANDUM

To: Gladys L. Martinez
From: Anna Martin
Date: January 1, 198-
Subject: Purchase Orders

You requested the number and the date of purchase orders for merchandise in transit. We have three:

    Purchase Order 2104, dated December 11
    Purchase Order 2103, dated December 10
    Purchase Order 2105, dated December 12

                    AM

    lm

MEMORANDUM

To: Calvin Schneider
From: Kiley Presson
Date: January 1, 198-
Subject: Appointments

Please add these appointments to my calendar:

    10:30 a.m., January 13, Cheryl Mathison
    11:30 a.m., January 10, Timothy Wong
    10:30 a.m., January 11, Glenda Thompson
    12:30 p.m., January 11, Katheryn Pappas
    10:30 a.m., January 12, Gregg Stilwell

Thanks.

                    KP

    lm

NAME ━━━━━━━━━━━━━━━━━━━━━━━━━━━━━━━━━━━ DATE ━━━━━━━━

MEMORANDUM

To: Janet Dewey and Stan Burns
From: Edna Suarez
Date: January 1, 198-
Subject: Suggestions from Employees

Three employees made suggestions this month: Dave
Gianni, Jenny Cantrell, and Steve Turner. Each
suggestion has merit.

Turner suggested staggering the arrival and departure
times of shift personnel. He feels this would
improve traffic congestion.

Cantrell recommended flexible working hours. She
included several articles that related successful
flex-time programs.

Gianni suggested a fitness trail. He says that
exercise will improve employee attitude, health,
and performance.

Please review these suggestions and decide which
one deserves our monthly cash award.

                              ES

lm
cc: Larson, Garner, and Rosenberg

NAME ▬▬▬▬▬▬▬▬▬▬▬▬▬▬▬▬▬▬▬▬▬▬ DATE ▬▬▬▬▬▬▬▬

REVIEW EXERCISE 10-B

Two weeks before an appointment, The Dental Clinic mails a reminder to each patient. The reminder, which is perforated in the middle, is a double postcard. The top part is a reminder, and the bottom part is a patient reply form.

Your dental appointment with _____

is _____ , _____ , at _____

Let us know if this appointment fits your schedule.

Simply detach the bottom portion of this card, complete it, and mail it. The card is addressed, and the postage is paid.

Keep the top portion for your reminder.

Dentist

The Dental Clinic

1916 Parker Road

Springfield, MA 01109

_____
Patient

_____
Address

_____
City, State ZIP

Please check one:

____ _____ , _____ , at _____ fits
       Day   Date   Time
my schedule.

____   Please call me at (   )   –
       to schedule another appointment.

_____ Signature

_____
Patient

The ten dentists at the Clinic insist on making their communications personal. To accomplish this, you have developed a form message with variables. **Variables** *are the parts of the message that are different for each addressee. The appointment clerk completes the variables form, and the word processing operator enters the variables and prints the cards.*

Yesterday, the Clinic purchased the software to handle the reminders. As office manager, you want to be sure that your new operator has mastered the software and that the software is operating as it should.

Part A

Assuming that the variables forms are correct, mark any inconsistencies between the forms and the printed cards, which appear on the next three pages.

Part B

Answer the following questions.

1. How many inconsistencies did you find?

_____

2. Do you think that the software is operating properly?

_____

3. Do you think that the new operator has mastered the software?

_____

NAME ▬▬▬▬▬▬▬▬▬▬▬▬▬▬▬▬▬▬▬▬▬▬ DATE ▬▬▬▬

VARIABLES FORM

Reminder:

Surname of dentist: _Dr. Barton_
Appointment day: _Tuesday_
Appointment date: _February 18_
Appointment time: _10:30 a.m._

Reply Address:

Dentist: _Dr. Elizabeth Barton_

Reminder Address:

Patient: _Mrs. Lynne Ortega_
Address: _Post Office Box 447_
City, State ZIP: _Springfield, MA 01109_

Reply:

Use day, date, and time variables above.

NO POSTAGE
NECESSARY
IF MAILED
IN THE
UNITED STATES

Mrs. Lynne Ortega
Post Office Box 447
Springfield, MA 01109

Please check one:

_____   Tuesday, February 18, at 10:30 a.m., fits
          my schedule.

_____   Please call me at (   ) -
          to schedule another appointment.

                    _____ Signature
          Lynne Ortega

Your dental appointment with Dr. Barton
is Tuesday, February 18, at 10:30 a.m.

Let us know if this appointment fits your schedule.
Simply detach the bottom portion of this card, complete
it, and mail it. The card is addressed, and the postage
is paid.

Keep the top portion for your reminder.

                    Lyn Davis

NO POSTAGE
NECESSARY
IF MAILED
IN THE
UNITED STATES

Dr. Elizabeth Barton
The Dental Clinic
1916 Parker Road
Springfield, MA 01109

NAME ▬▬▬▬▬▬▬▬▬▬▬▬▬▬▬▬▬▬▬▬▬▬▬▬▬▬ **DATE** ▬▬▬▬▬▬▬▬▬

VARIABLES FORM

Reminder:

Surname of dentist: *Dr. Fenton*
Appointment day: *Thursday*
Appointment date: *February 20*
Appointment time: *11:30 a.m.*

Reply Address:

Dentist: *Dr. J.L. Fenton*

Reminder Address:

Patient: *Mr. Adam Crosby*
Address: *3534 Donald Avenue*
City, State ZIP: *Springfield, MA 01107*

Reply:

Use day, date, and time variables above.

Mr. Adam Crosby
3534 Donald Avenue
Springfield, MA 01107

Please check one:

_____ Thursday, February 20, at 11:30 a.m., fits my schedule.

_____ Please call me at (   ) - to schedule another appointment.

_____ Signature
Adam Crosby

Your dental appointment with Dr. Fenton is Thursday, February 20, at 11:30 a.m.

Let us know if this appointment fits your schedule. Simply detach the bottom portion of this card, complete it, and mail it. The card is addressed, and the postage is paid.

Keep the top portion for your reminder.

Lyn Davis

Dr. J. L. Fenton
The Dental Clinic
1916 Parker Road
Springfield, MA 01109

NAME _____ DATE _____

VARIABLES FORM

Reminder:

Surname of dentist: *Dr. Mc Gee*
Appointment day: *Friday*
Appointment date: *February 21*
Appointment time: *9:30 a.m.*

Reply Address:

Dentist: *Dr. Kenneth Mc Gee*

Reminder Address:

Patient: *Ms. Martha Irvin*
Address: *Post Office Box 3815*
City, State ZIP: *Springfield, MA 01101*

Reply:

Use day, date, and time variables above.

NO POSTAGE
NECESSARY
IF MAILED
IN THE
UNITED STATES

Ms. Martha Irvin
Post Office Box 3815
Springfield, MA 01101

Please check one:

_____   Friday, February 20, at 9:30 a.m., fits
          my schedule.

_____   Please call me at ( ) -
          to schedule another appointment.

_____ Signature

Martha Irvin

Your dental appointment with Dr. McGee
is Friday, February 20, at 9:30 a.m.

Let us know if this appointment fits your schedule.
Simply detach the bottom portion of this card, complete
it, and mail it. The card is addressed, and the postage
is paid.

Keep the top portion for your reminder.

Zayn Davis

NO POSTAGE
NECESSARY
IF MAILED
IN THE
UNITED STATES

Dr. Kenneth McGee
The Dental Clinic
1916 Parker Road
Springfield, MA 01109

NAME ▬▬▬▬▬▬▬▬▬▬▬▬▬▬▬▬▬▬▬▬▬▬▬   DATE ▬▬▬▬

THE FINISHED PRODUCT

CHAPTER

Editing and Proofreading

The Final Copy

Suppose a document has been edited to be correct, complete, clear, courteous, concise, and consistent. The revised document is then prepared in one of several ways, depending on the equipment, the original form, and the magnitude of the changes.

TRADITIONAL EQUIPMENT

Minor corrections or changes can be made directly on a typed document. However, some documents must be typed or retyped because (1) the original is handwritten, (2) the original is in draft form, or (3) the number and kind of revisions make retyping necessary.

AUTOMATED EQUIPMENT

All changes can be made on the screen, and a correction-free document can be printed.

CHECKING CORRECTIONS

Now that the document has been corrected, typed, or printed, is the editing process finished? No! The revised document should be proofread against the original edited version to see if (1) *all* revisions were made, (2) revisions were *correctly* made, and (3) *new errors* were made.

The next few exercises will give you practice in proofreading the edited original with the revised document.

EXERCISE 11-1

Is each paragraph a finished product? Compare the edited copy with the final copy. Use revision symbols to mark any revisions that were overlooked or not correctly made. Also, mark any new errors. If the final copy is correctly revised, write OK beside the item number.

1. Raymond wrote an up-to-date report on computer usage. His report includes several recommendations for relieving the problem areas.

____ 1. Raymond wrote an up-to-date report on computer usage. His report includes several reccomendations for relieving the problem areas.

2. Wilburn has studied up on Cobol and Fortran. Most of his experience has been with IBM COMPUTERS.

____ 2. Wilburn has studied COBOL and Fortran. Most of his experience has been with IBM Computers.

3. Lavonda has has very good human relations skills. Her rapport with subordinates seems to be based on mutual trust and respect.

____ 3. Lavonda has very good human relations skills. Her rapport with subordinates seems to be based on mutual trust and respect.

4. Pat is a expert on business writing. Many companies ask her to help them write effective sales and collection letters. Other companies ask her to conduct in-house clinics to improve the writing skills of their executives.

____ 4. Pat is a expert on business writing. Many companies ask her to help them write effective sales and collection letters. Other companies ask her to conduct in-house clinics to improve the writing skills of their execatives.

NAME ══════════════════════════════ DATE ══════════════

EXERCISE 11-2

Complete Parts A, B, and C.

Part A

The report draft below is based on a survey completed by the members of the City of Springs Chapter of Professional Secretaries International. Compare the draft and the revised copy, marking any needed revisions.

A
EXPERIENCE AND CPS RATING
⊐ A ⊏   *on half sheet*

of the
City of Springs Chapter
Professional Secretaries International
Springs, North Carolina

⊐August 30, 198- ⊏

Number of Secretaries	Years of Experience	CPS Rating
B 12	15 and over	10
8	10–14	5
10	6–9	3
15	4–5	7
6	0–3	2 B
~~50~~ 51		27 A

B ds

EXPERIENCE AND CPS RATING

of the
City of Springs Chapter
Professional Secretaries International
Springs, North Carolina

August 30, 198-

Number of Secretaries	Years of Experience	CPS Rating
12	15 and over	10
8	10–14	5
10	6–9	3
15	4–5	7
6	0–3	2
50		27

NAME ▬▬▬▬▬▬▬▬▬▬▬▬▬ DATE ▬▬▬▬▬▬▬

Part B

After the report was edited and typed, several changes were needed to make the report accurate. Make the changes on the revised copy.

1. Two secretaries received their Certified Professional Secretary (CPS) rating in today's mail. One is in the 6–9 year category, and one is in the 4–5 year category.
2. Two new members completed the survey. Both are in the 10–14 year category; one is a CPS.
3. To get 100 percent participation, an absent member was surveyed by mail. This member is in the 10–14 year category and is a CPS.

Part C

Your instructor may ask you to type the final copy, including the changes you made, on a separate sheet of paper.

EXERCISE 11-3

Complete Parts A, B, and C.

Part A

Fran's note, which appears below on the left, was the basis for Amos Lorenzo's memorandum, on the right. After the memorandum was edited, Lee Harris decided to go on the tour and gave Mr. Lorenzo a check for $400. This is when Mr. Lorenzo realized that every incidence of $300 should be changed to $400.

Assuming that the note is correct, use revision symbols to mark all needed changes in the memorandum. Also look for errors that may have been missed in the original editing. Do not use global symbols.

NOTE

Mr. Lorenzo,

Below is the payment status for the New England tour.

Paid in Full

Jose Rodriques
Art Schaeffer
Barbara Tanner
Wendy Younce

Paid $200

David Bronson
Patrick Dillon
Susan Gaither
Ann Underwood
Sandra Ware

The last payment is due September 11.

Fran
8/25/8-

MEMORANDUM

To: George Donnelly
From: Amos Lorenzo
Date: August 25, 198-
Subject: Company Sponsored Trip

Below is a list of people who have paid part or all of the $300 cost for the fall bus tour through New England. Their last payment is due September 1.

Name	Amount
David Bronson	$ 200
Patrick Dillon	200
Susan Gaither	200
Jose Rodriques	300
Art Schaeffer	300
Barbara Tanner	300
Ann Underwood	200
Sandra Ware	200
Wendy Younce	300
Total Collected	$2,200

Bill Anders from the travel agency will meet with the (9) people tomorrow morning at 7:30 in Room 303. Please give him a check for the total cost of the tour. Make the check for $2,700\00 payable to Host Tours, Inc.

AL

sp

Part B

Below are two checks. One is complete except for the signature; the second one is blank. After finishing Part A, decide which of the two choices below is correct. Make your selection and follow the instructions.

Choice 1

If Check 1074 is correct:

1. *Sign the check (your name).*
2. *Write VOID in all-capital letters across Check 1075.*

Choice 2

If Check 1074 is incorrect.

1. *Write VOID in all-capital letters across Check 1074.*
2. *Write Check 1075 for the correct amount. Sign the check (your name).*

THE MANUFACTURING COMPANY
Post Office Box 8914
Jacksonville, Florida 32207

63-195
630

No. __1074__

August 26, 198-

Pay to the
Order of Host Tours, Inc.-------------------- $ 3,600.00

Three Thousand Six Hundred and 00/00----------------- Dollars

Trust Bank, Inc.
Jacksonville, Florida 32207

Authorized Signature

⑆0630⑉01951⑆ 346 ⑈2809⑈028

THE MANUFACTURING COMPANY
Post Office Box 8914
Jacksonville, Florida 32207

63-195
630

No. __1075__

198 ____

Pay to the
Order of _____ $ _____

_____ Dollars

Trust Bank, Inc.   —SPECIMEN CHECK ONLY—
Jacksonville, Florida 32207

Authorized Signature

⑆0630⑉01951⑆ 346 ⑈2809⑈028

Part C

Your instructor may ask you to type the corrected memorandum on a separate sheet of paper.

REVIEW EXERCISE 11-A

Part A

A résumé service prepared the draft for William Shapiro. Mr. Shapiro edited the draft. Compare the edited draft below with the printed résumé on page 145. Use revision symbols to mark any changes not made on the printed copy.

Part B

Your instructor may ask you to type the revised résumé, including your changes, on a separate sheet of paper.

WILLIAM T. SHAPIRO

Home Address   School Address

3746 Covall Drive   Post Office Box 849
Madison, WI 53713   Green Bay, WI 54305
608-555-8284   414-555-7890

SPORTS   Football, ~~Burton~~ Barton University   ⟝ more A
                    Baseball, Burton University

EDUCATION   Burton University, Green Bay, Wisconsin
                    B.S. in Accounting, Anticipated May 198-

                    City Community College, Madison, Wisconsin
                    A.A. in Business Administration, May 1983

                    Central High School, Madison, Wisconsin
                    Diploma in College Preparation Curriculum, June 1981

EXPERIENCE   Hamburger Delight, Green Bay, Wisconsin
                    Supervisor, part-time employees; 1981-present

                    Jones Construction, Inc., Madison, Wisconsin
                    Carpenter's helper; 1979-1980

Awards   Senior Accounting Award, Burton University
                    Manson Citizenship Award, Burton University
                    Phi Kappa Phi Honor Society, Burton University

ORGANIZATIONS   Fellowship of Christian Athletes
                    Phi Beta Lambda, City Community College and Burton
                    Future Business Leaders of America, Central High School

POSITION   My objective is an entry-level accounting position that more
OBJECTIVE   offers challenge and advancement.   B

REFERENCES AND   References and transcripts are available on request.
TRANSCRIPTS

WILLIAM T. SHAPIRO

Home Address

3746 Covall Drive
Madison, WI 53713
608-555-8384

School Address

Post Office Box 849
Green Bay, WI 54305
414-555-7890

POSITION
OBJECTIVE

My objective is an entry-level accounting position that
offers challenge and advancement.

EDUCATION

Barton University, Green Bay, Wisconsin
B.S. in Accounting, Anticipated May 198-

City Community College, Madison, Wisconsin
A.A. in Business Administration, May 1983

Central High School, Madison, Wisconsin
Diploma in College Preparation Curriculum, June 1981

EXPERIENCE

Hamburger Delight, Green Bay, Wisconsin
Supervisor, part-time employees; 1981-present

Jones Construction, Inc., Madison, Wisconsin
Carpenter's helper; 1979-1980

AWARDS

Senior Accounting Award, Barton University
Manson Citizenship Award, Barton University
Phi Kappa Phi Honor Society, Barton University

ORGANIZATIONS

Fellowship of Christian Athletes
Phi Beta Lambda, City Community College and Barton
Future Business Leaders of America, Central High School

SPORTS

Football, Barton University
Baseball, Barton University

POSITION
OBJECTIVE

My objective is an entry-level accounting position that
offers challenge and advancement.

REFERENCES AND
TRANSCRIPTS

References and transcripts are available on request.

REVIEW EXERCISE 11-B

Part A

*The use of **boilerplate** paragraphs (often called **form** paragraphs) is a time-saver when the same information is written repeatedly in business messages. Instead of dictating the same information each time, the business person simply chooses which boilerplate paragraphs are to be used and gives the new address information.*

    Below is the first draft of boilerplate paragraphs written by your assistant. Because it is the first draft, the paragraphs can be improved significantly. Use revision symbols to edit the paragraphs.

Paragraph 1

Thank you for requesting and asking for travel information.

Your request gives us teh opportunity to tell you how our experienced staff can help plan your trip.

Paragraph 2

Enclosed is the Carribbean cruise brochure you wanted. Each cruise leaves from Miami, however, affordable air transportation to Miami can be added to your travel package though our arrangment with cooperating airlines.

Paragraph 3

Enclosed is the European tour brochure you wanted. You may chose from several itineraries and departure dates. All tours have planed activitys and free time.

Paragraph 4

Enclosedis the canadian motor tour brochure you wanted. You will fly directly from New Yorks Kennedy Air Port to Montreal. Where your tour begins. If you are afraid to fly, you can choose form several scenic bus tours and meat your group in Canada.

Paragraph 5

As a Citizen of the United States, you will not need a passport to visit Canada. You will need, however a official copy of our birth certificate.

Paragraph 6

Eventhough your plans are indefinite, you should apply for for your passport now. The Main Street Post Office provide this service each Tuesday afternoon from 1:30 to 4:30 a.m. You should get your passport in 4 to 8 weeks after applying.

Paragraph 7

Unless your passport has been renewed sense last traveling with us, it will expire before our next tours begin. Please check your passport and renew it if necessary.

Paragraph 8

Each traveler in our organized tours are insured by Travel Insurance, Inc. This coverage is provided as a service and is included in the tour price.

Paragraph 9

Your traveling pleasure is our goal. To help you enjoy your trip and relieving you of time-consuming details, your group will have a guide thoughout the trip.

Paragraph 10

Please do not hesitate to call us up if you have questions or can't understand the information in the brochure. If you make reservations with in 15 days the of date of this letter, you will get a 5 percent discount. Call us today to plan your trip!

Part B

Your instructor may ask you to type the revised paragraphs on a separate sheet of paper.

Part C

Sean Trevor, the manager of Hampton Travels, Inc., used the revised paragraphs to answer four telephone inquiries. He made notes during the phone calls and added the paragraph numbers later. Are the correct paragraphs used on the form letter requests that appear on page 148? Strike through an incorrect paragraph number and write the correct number beside it.

Part D

Your instructor may ask you to type the letters on separate sheets of paper using the revised paragraphs and the corrected letter request forms. Use modified-block style, standard punctuation, indented paragraphs, and your reference initials. After completing each letter, proofread it by comparing it to the original copy. Sign and date the letter request form.

NAME ▬▬▬▬▬▬▬▬▬▬▬▬▬▬▬▬▬ DATE ▬▬▬▬▬▬▬

REQUEST FOR FORM LETTER

S. Trevor _____ _2/24/8-_
Requested By   Date

TO _Mr. Daniel Benitez_
Tudor Industries
Post Office Box 2987
Greenville, SC 29605

Notes	Paragraphs
Thank you	_1_
Europe	_3_
Apply passport	_6_
Discount	_10_

Typed By _____   Date _____

REQUEST FOR FORM LETTER

S. Trevor _____ _2/24/8-_
Requested By   Date

TO _Mrs. Sherry Sawyer_
1812 Camden Lane
Greenville, SC 29605

Notes	Paragraphs
Thank you	_1_
Canada	_4_
Birth certificate	_5_
Guide	_9_
Discount	_10_

Typed By _____   Date _____

REQUEST FOR FORM LETTER

S. Trevor _____ _2/24/8-_
Requested By   Date

TO _Mr. and Mrs. Carl White_
113 Gentry Circle
Greenville, SC 29611

Notes	Paragraphs
Thank you	_1_
Caribbean	_2_
Insurance	_9_
Discount	_10_

Typed By _____   Date _____

REQUEST FOR FORM LETTER

S. Trevor _____ _2/24/8-_
Requested By   Date

TO _Ms. Michelle Wey_
1901 Hilton Street
Greenville, SC 29607

Notes	Paragraphs
Thank you	_1_
Europe	_2_
Renew passport	_7_
Discount	_10_

Typed By _____   Date _____

NAME ═══════════════════   DATE ═══════════════

CHAPTER

12

Applied Editing

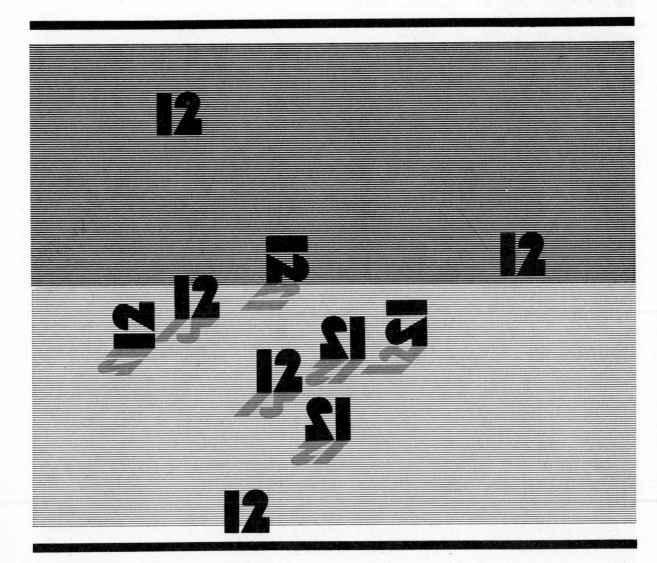

On the Job

This chapter contains an editing application like one you might find on the job. The application is made up of editing tasks related to an office automation seminar.

You will use much of what you have learned from your instructor and from this book. You will also see how one change makes other changes necessary. For example, if the seminar date is changed, you must correct the program and all memos and letters.

REVIEW EXERCISE 12-A

You work for a management consulting firm that offers seminars on business topics. As assistant to B. J. Arnold, company president, you help with seminar arrangements.

Today is May 5. You have just returned from a business trip to find the items listed below on your desk. The items are related to an office automation seminar your company is sponsoring. (See pages 151 to 156.)

Read all the instructions and communications before beginning. Use revision symbols to edit the communications that you are responsible for sending inside or outside your company. Be sure to check the dates of the communications because the order of events may affect the revisions. Check a reference source when you are uncertain about mechanics such as grammar, spelling, and punctuation.

Note from B. J. Arnold	*Requests for copies (2)*
Memorandum with an attachment	*Telephone message*
Reservation Confirmation from	*Request for check*
the Conference Center	*Registration forms (2)*
Envelope addressed to Conference Center	

You should be familiar with these facts: (1) The seminar is May 13 to 15 at the City College Conference Center. (2) Cost for the seminar is $500 per person. (3) You must request refund checks for all cancellations. (4) B. J. Arnold has specified that the number of copies to be made for seminar items must be the final number of registered participants plus 5.

REVIEW EXERCISE 12-B

On pages 157, 159, and 161, you will find the final typed copy and the artwork that must be assembled to make up the program for the office automation seminar. It is your job to make up the two master sheets needed for the process.

To see what the finished, four-page program booklet will look like:

1. Take an 8½- by 11-inch sheet of unruled paper. Hold the sheet so that the 11-inch side is horizontal.

2. Now fold the sheet down the middle of the 11-inch side. You have a small, four-page booklet measuring 5½- by 8-inches.

3. Write the number 1 on the first page. Open the booklet and number pages 2 and 3. Turn the last page and write the number 4.

The masters for the booklet will be printed back-to-back on the same sheet of paper. You will need two masters—one for each side of the printed sheet. To see how the masters must be set up in order to achieve the booklet result:

1. Open the booklet you made so that pages 1 and 4 are facing you. Notice that page 1 (title page) is on the right and page 4 is on the left. Follow this arrangement to create Master Sheet 1.

2. Now turn the booklet over so that pages 2 and 3 are facing you. Notice that page 2 is at the left and page 3 is at the right. Follow this arrangement to create Master Sheet 2.

Now you are ready to create the masters:

1. Take two sheets of 8½- by 11-inch unruled paper. Lightly draw a line down the center of each sheet and label the parts of the sheets with page numbers. (These pencil marks can be erased or covered after the sheets are finished.)

2. Start with the title page. Cut out the material that goes on this page and tape or glue it down.

3. Make sure to follow these instructions:

   a. The drawing of a computer, which appears on the bottom of page 161, goes on the title page.

   b. The footnotes are marked with asterisks (*). They must go at the bottom of the appropriate pages.

   c. Remember that the door prize registration form will be detached from the program. Make sure that the program listing will remain complete after the form is detached.

4. Continue with pages 2, 3, and 4 in order. Try to balance the material on the pages so that the pages look attractive.

Note From B. J. Arnold

NOTE

As you know, I'll be in Dallas until May 8. Please check my mail and messages for seminar registration changes.

Even though the deadline is passed, let's accept registrations received by May 5. Hold all communications until May 6 so that the final figures can be used.

Please check the memo carefully. I haven't worked with this secretary previously and must leave before the memo is transcribed.

Thanks,
B. J.
5/1

NAME ━━━━━━━━━━━━━━━━━━━━━━━━━━━ DATE ━━━━━━━

Memorandum With Attachment

MEMORANDUM

To: Fran Murray, Seminar Leader
From: B. J. Arnold, Manager
Date: April 31, 198-
Subject: Office Automation Seminar--May 13-15

Sixty-five people from ten companies are registered for the office
automation seminar. A list of the companies and the number attending
from each one are attached. I have requested copys of your handouts,
and my assistant will confirm the reservations at the conference
center

Good luck on your 1st seminar! I plan to attend so I can point out
your errors at our follow-up evaluation. Please let me know if I can
can help you in anyway.

                         JB

rh
Enclosure
cc: Adams, Bronson, Reynolds, and Davis

OFFICE AUTOMATION SEMINAR
May 13-15, 198-

Participating Companies	Number of Participants
Askew Manufacturing	10
Cansler, Inc.	3
City Hospital	5
Court Reporters, Inc.	2
Frieght Carriers, Inc.	10
Harbison Sales Company	15
Johnson Industries	5
Kelco Print Shop	5
Rosen Import Company	7
Whitney Temporaries	3
Total	65

NAME ▬▬▬▬▬▬▬▬▬▬▬▬▬▬▬▬▬▬▬▬▬▬▬▬ DATE ▬▬▬▬

Reservation Confirmation

RESERVATION CONFIRMATION

Reservations for your group are confirmed as follows:

Name: B.J. Arnold, President
Company: Management Consultants, Inc.
              Post Office Box 8271
              Asheville, NC 28804

Number of Rooms: _____75_____   Dates: __May 13–14__

Please complete the items below. Keep the original and mail the copy to me by May 8.

Thank you,

Steve White
Steve White

— —

Is the number of rooms correct? _____

If not, how many rooms do you need? _____

Signed: _____   Date: _____

Envelope

**MANAGEMENT
CONSULTANTS, INC.**
Post Office Box 8271
Asheville, NC 28804

Mr. Steve White
City College conference Center
Post Office Box 2431
Asheville, NC 2880

Requests for Copies

REQUEST FOR COPIES

Date needed   5/25

Title   Office Automation Program

File number   Will be sent to you 5/6

Number of copies   70

Requested by   B. J. Arnold

REQUEST FOR COPIES

Date needed   5/25

Title   Office Automation Hand-Outs

File number   5523

Number of copies   70

Requested by   B. J. Arnold

Telephone Message

```
For B.J. Arnold
Date 5/3              Time  10:35

WHILE YOU WERE OUT
Mr. Keith Barnett
From Cansler, Inc.
     1831 Cedar Cliff Road
     Asheville, NC  28803
Phone No. (704) 555-2163
         Area Code    Number    Extension
```

TELEPHONED	✓	URGENT	
PLEASE CALL		WANTS TO SEE YOU	
WILL CALL AGAIN		CAME TO SEE YOU	
RETURNED YOUR CALL			

```
Message Cansler has three
reservations for the office
automation seminar. Please
cancel one.
                    A. Faison
                         Operator
```

Request for Check

```
              REQUEST FOR CHECK

                        $ _____
                           Amount


Payable to
_____
_____
_____
_____

Reason
_____
_____
_____

Approved by _____

Date _____
```

NAME ━━━━━━━━━━━━━━━━━━━━━━━━━━━━━━━━━━━━━ DATE ━━━━━━━━━━━━

Registration Forms

REGISTRATION FORM

Office Automation Seminar

May 13-15, 198-

RECEIVED
5/3/8-

Participants

Samuel A. Berman

Peter Koffman

Lisa Wexler

Company name and address

Zennon Insurance Company, Inc.

5814 Kenwood Drive

Asheville, NC 28806

Number attending ___3___ x $500 = $1,500

Check is enclosed (X)   <u>or</u>   Please invoice company (   )

REGISTRATION FORM

Office Automation Seminar

May 13-15, 19--

RECEIVED
5/4/8-

Participants

Laura A. Delgado

_____

_____

Company name and address

First Public Bank

Post Office Box 3219

Asheville, NC 28802

Number attending ___1___ x $500 = $500

Check is enclosed (X)   <u>or</u>   Please invoice company (   )

NAME ■■■■■■■■■■■■■■■■■■■■■■■   DATE ■■■■■■■■■■

O F F I C E   A U T O M A T I O N

<u>A SEMINAR FOR PROFESSIONALS</u>

May 13-15, 198-

CITY COLLEGE CONFERENCE CENTER
ASHEVILLE, NORTH CAROLINA

Sponsored By
Management Consultants, Inc.
Post Office Box 8271
Asheville, North Carolina 28804

OFFICE AUTOMATION SEMINAR
Thursday, May 13, 198-

8:30-9:00	Continental Breakfast	Room 103
9:00-10:30	Office Trends Dr. Allison Barkley, Author of Best-Seller: <u>High Tech Trends</u>	Room 105
10:30-11:00	Coffee Break	Room 103
11:00-12:00	Information Systems	Room 106
12:00-1:30	Lunch	Cafeteria
1:30-3:00	Employee Development College Courses In-House Courses	Room 201
3:00-3:30	Coffee Break	Room 103
3:30-6:00	Industry Tour Informations Center Star Industries, Inc.	Lobby*
7:00-8:30	Dinner	Dining Room

NAME ━━━━━━━━━━━━━━━━━━━━━━━━━━━━━━━━ DATE ━━━━━━━━━━

Friday, May 14, 198-

8:30-9:00	Continental Breakfast	Room 103
9:00-10:30	Word Processing Concepts	Room 106
10:30-11:00	Coffee Break	Room 103
11:00-12:00	Word Processing Skills	Room 104
12:00-1:30	Lunch	Cafeteria
1:30-3:00	Word Processing Equipment	Room 208
3:00-3:30	Coffee Break	Room 103
3:30-6:00	Equipment Demonstrations	Room 218
7:00-8:30	Dinner*	Dining Room

DOOR PRIZE

OFFICIAL REGISTRATION FORM

Name _____

Company Name _____

_____

Companies Donating Door Prizes

Calhoun Computer Center

Mostello's Department Store

Outlet Mall, Inc.

_____

NAME ━━━━━━━━━━━━━━━━━━━━━━━━━━━━   DATE ━━━━━━━━━━━

Saturday, May 15, 198-

8:30-9:00	Continental Breakfast	Room 103
9:00-10:30	Telecommunications Award-Winning Movie: <u>Making Connections</u>	Room 310
10:30-11:00	Coffee Break	Room 103
11:00-12:00	Reprographics Micrographics	Room 318
12:00-1:00	Lunch*	Cafeteria
2:00	Check-Out Time	

*Please meet in the lobby at 3:30 p.m. Transportation will be provided.

*You must be present at Saturday lunch to win door prize.

*Register for door prizes Friday night at dinner. The drawing will be Saturday at lunch.

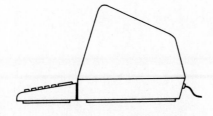

NAME ▬▬▬▬▬▬▬▬▬▬▬▬▬▬▬▬▬▬▬▬▬▬▬ DATE ▬▬▬▬▬

REFERENCE SECTION

I. CAPITALIZATION

1. Capitalize the first word in a sentence.

   Give the report to the manager.

2. Capitalize the first word of a direct quotation.

   Fred said, "Please call me."

3. Capitalize the names of particular people, places, and things.

   Mrs. Joy Ramsey, from Orange Park, Florida, uses Teflon pans.

4. Do not capitalize common nouns.

   A customer ordered two dozen pans.

5. In a salutation of a letter, capitalize the first word, all nouns, and all official titles.

   Dear Mr. Parker:
   Ladies and Gentlemen:
   Dear Mrs. Sands and Professor Murphy:
   Dear President Flanders:

6. Capitalize only the first word in the complimentary closing of a letter.

   Sincerely,
   Sincerely yours,
   Very truly yours,

7. Capitalize the days of the week, months of the year, and holidays. Seasons are *not* capitalized.

   Our only spring holiday is Easter.
   The office will be closed Friday, April 3.

8. Capitalize titles when they are part of a person's name. Do not capitalize titles when they follow or further explain a person's name.

   Tom Farmer, visiting professor, asked Professor Tim Ryan.

9. Do *not* capitalize *a.m.* or *p.m.*.

   The cafeteria is open every day from 11:15 a.m. until 2:30 p.m.

10. Use either all capitals or underscores for book titles, but be consistent within a communication.

    All Capitals
    He read MATH MADE EASY.

    Initial Capitals with Underscores
    He read Math For Beginners.

    Within the Same Communication:
    He read MATH MADE EASY and MATH FOR BE-GINNERS.

    OR

    He read Math Made Easy and Math For Beginners.

11. In article titles, use initial capital for the first word and all other words that are not articles or prepositions. (Article titles are enclosed in quotation marks.)

    The title of her article is "How to Set Priorities."

12. Use all capitals for coined acronymns and certain abbreviations. (When in doubt, check a dictionary.)

    She studied BASIC, COBOL, and FORTRAN.
    Please give us your ZIP code.

13. Capitalize points of the compass (or their derivatives) when they are used to specify a definite region. Do *not* capitalize them when they are used only as direction or general location.

    They plan to open a travel bureau on the West Coast. (Definite region)
    The best site is east of the railroad. (Direction)
    They chose a city in the Deep South. (Derivative used as definite region)
    He wants to live on the west side of town. (General location)

14. Use all capitals for two-letter state abbreviations used in addresses.

    Dr. Steven West
    143 Third Street
    Chicago, IL 60612

15. In a memorandum, use all capitals for the title. Use initial capitals for the guide words and all other words that are not articles or prepositions.

MEMORANDUM

To:	Henry Rankin
From:	Hilda Polski
Date:	June 15, 19—
Subject:	Sales for April and May

16. Use these rules for reports:
    a. For titles and side headings use all capitals.
    b. In subtitles or by-lines, use initial capital for the first word and all other words that are not articles or prepositions.
    c. In paragraph headings, use initial capital for the first word and all other words that are not articles or prepositions. Paragraph headings are underscored.

INCREASING OUR PROFIT
By Shirley Linnen

REDUCING WASTE

<u>Quality of Materials.</u> Buying cheaper isn't always buying better. Studies . . .

II. CONSISTENCY

Similar items should be treated consistently. Here are some examples:

1. Treat similar parts of a sentence the same. (See "Parallelism" in Section III.)

2. Subjects should agree with their verbs and pronouns should agree with their antecedents. (See "Subject-Verb Agreement" and "Pronoun-Antecedent Agreement" in Section III.)

3. Within a communication, use all capitals for all book titles or underline all book titles. Both forms are correct, but treatment should be consistent.

4. Even though two different courtesy titles may apply to the same person—such as *Ms.* and *Dr.*—use one title throughout the same document.

5. Treat names the same. When listing both men's and women's names, use a courtesy title for *all* or *none* of the names.

6. Do not mix format styles within a document. For example, do not indent one paragraph in a letter when all other paragraphs are not indented.

7. In the same sentence, express related numbers in all words or all figures.

    two cans and four jars   **NOT**   *two cans and 4 jars* (See Section IV.)

8. Either abbreviate or spell out terms. For example, do not use *gallons* and *gals.* in the same document.

9. Check for correct spelling of names and make sure the correct spelling is used in each instance. For example, which is correct, *McDonald's* or *Mac-Donald's*?

10. There are some cases where the use of commas is a matter of style. For example, *Will Lutz, Jr.* and *Will Lutz Jr.* The commas here are neither "right" nor "wrong" but a matter of individual preference. (See "Commas" in Section V.)

11. Statements must be consistent with known facts and source documents. It is sometimes necessary to check a reliable source.

    Our club meets the second Tuesday of each week. (Should be *of each month.*)
    The report is due Monday, May 3. (Verify that Monday is the third of May.)
    The bill yesterday was $200 and the bill today was $200. That makes a total of $300 to repair the furnace. (The total should be $400.)

12. Do not omit the first or last part of paired punctuation marks. "You're hired! should be "You're hired<u>!</u>" and (1 should be and (1).

13. In enumerations and lists use one number or letter style.

    NOT
1.	A.
2.	b.
(3.)	C.

14. Choose the most appropriate sequence with related items. A special sequence may be specified, but the most commonly used are alphabetical, chronological, and numerical orders.
    a. Use alphabetical order when mentioning names within a communication or when listing names of people receiving copies. This reduces unintentional bias.
    b. Use chronological order when the time sequence of events is important.
    c. Use numerical order when listing invoice numbers, check numbers, etc.

III. GRAMMAR
Sentence Fragments

1. A complete sentence has a subject and a verb.

    Our manager approved the expense. (Subject, *manager*; verb, *approved*)
    Mrs. Haley and Mr. Roe are in the conference room. (Subjects, *Mrs. Haley and Mr. Roe*; verb, *are*)
    Please make another copy for me. (Subject is understood to be *you—you* make . . .)

2. When a subject or a verb is missing, the group of words is a fragment.

    Early in the morning last Tuesday. (No verb)

3. Some clauses may have a subject and a verb but may still be incomplete.

    When our supervisor arrives. (Fragment)
    Because she was late this morning. (Fragment)

What will happen "When our supervisor arrives"? What happened "Because she was late this morning"? These fragments do not tell us. They are incomplete.

    When our supervisor arrives, we will discuss this schedule. (Sentence)
    Because she was late this morning, she missed the boss's announcement. (Sentence)

Subject-Verb Agreement

1. Subjects and verbs must agree. A singular subject must have a singular verb; a plural subject must have a plural verb.

    My assistant is on vacation. (*Assistant*, the subject, is singular; *is*, the verb, is singular. They agree.)
    My assistants are on vacation. (Here, *assistants* and *are* agree. Both are plural.)

2. Subjects joined by *and* are plural. There are two subjects.

> Mrs. Winn and Mr. Klee are here. (Both are here.)

3. Two singular subjects joined by *or* need a singular verb.

> Mrs. Winn or Mr. Klee is here. (Only one is here.)

4. When a singular and a plural subject are joined by *or*, make the verb agree with the subject closer to the verb.

> Mrs. Winn or the managers have this report.
> The managers or Mrs. Winn has the report.

Pronoun-Antecedent Agreement

Because pronouns *replace* nouns, pronouns must agree with the nouns that they replace. In other words, they must "follow the leader." They must agree in number and in gender with the leader noun:

a. The *man* who was here earlier said *he* wanted to see Miss Jenkins.
b. The *woman* who was here earlier said *she* wanted to see Miss Jenkins.
c. The *people* who were here earlier said *they* wanted to see Miss Jenkins.
d. An *employee* must show *his or her* identification card to the guard.
e. *Employees* must show *their* identification cards to the guard.

Parallelism

Parallelism simply means treating similar constructions in a similar manner. Adjectives should be paralleled by adjectives, nouns by nouns, infinitives by infinitives, subordinate clauses by subordinate clauses, etc. In the examples, note especially the words in italics.

Not Parallel	**Parallel**
She is both *skilled* and *has experience*.	She is both *skilled* and *experienced*.
My duties include *scheduling* and *estimates*.	My duties include *scheduling* and *estimating*.
His hobbies are *playing* tennis and *to cook* gourmet meals.	His hobbies are *playing* tennis and *cooking* gourmet meals.
She approved not only *my raise* but also *promoting me*.	She approved not only *my raise* but also *my promotion*.
Give copies both *to Mr. Cohn* and *Mrs. Dent*.	Give copies both *to Mr. Cohn* and *to Mrs. Dent*.

IV. NUMBERS

1. Verify math calculations in content, on invoices, in reports, in tables, etc.
   Content

   > *Three* of the *five* forms have been returned. We hope to get the other *two* by Monday.

2. In general, spell out numbers from 1 through 10, and use figures for numbers above 10, However, be consistent in using *related* numbers within a sentence. Also, spell out a number that begins a sentence. Reword the sentence if it is awkward to spell out the number.

Numbers One Through Ten

All three employees were promoted.

Numbers Above Ten

Only 15 people voted against the motion.

Numbers Above and Below Ten

One manager, two assistant managers, and fifteen supervisors liked the plan. (All were spelled out for consistency.)

Related and Unrelated Numbers

Only 5 of the 20 people attended all three sessions. (Figures are used for 5 and 20 because both are *related* to people. The number of sessions, *three*, is not related to the number of people and is spelled out because it is under ten.)

3. Use figures for amounts of money, technical measurements and specifications, percentages, and fractions (unless the figure begins the sentence).

10.5 cm	$1,223.45	65 cents
2.2 percent	2.4 feet	4:1 ratio
3.5%	12⅞	12¼

   a. Use the word *percent* except in statistical writing or in columns where space is restricted. In these cases, use the percent symbol (%).
   b. Be consistent in using constructed fractions and symbol fractions.

   > The room is 22¼ feet by 12¾ feet.
   > The room is 22 1/4 feet by 12 1/8 feet.

4. Use the word *cents* with amounts under a dollar. Use the symbol ¢ in statistical writing or in columns when space is restricted. Within a sentence, omit .00 in whole dollar amounts. However, in columns, add .00 if any number has cents.

   I owe him 80 cents.
   The mechanic charged $69 for the repairs.
   The first invoice was $399.50, and the second invoice was $100.

$399.50
100.00
$499.50

5. Use figures to express time with the expression *o'clock* and with *a.m.* or *p.m..* Omit *:00* for time "on the hour."

   The meeting will end by 5 o'clock.
   We open at 7 a.m.
   We open at 6 a.m. and close at 11:30 p.m.

6. Spell out ages except in technical writing or when used as a significant statistic.

   Technical writing
   Our records show that most people retire between the ages of 62 and 70.

   Significant statistic
   Bill, at 21, is already a millionaire.

   Nontechnical and nonstatistical
   Minimum employment age is eighteen.
   (Spell out indefinite numbers or amounts and fractions that stand alone.)
   About a thousand people attended.
   New equipment will cost thousands of dollars.
   The motion passed with a two-thirds majority.

7. Use figures in dates as follows:
   We began the project on May 1, 1983.

We began the project on the first of May.
On July 8, we started making a profit.

8. Use Roman numerals for chapters of a report and for parts of an outline.

OUTLINE

   I. Purchasing the Equipment
   II. Leasing the Equipment
   III. Recommendations

9. Ordinal numbers are usually spelled out if they can be expressed in one or two words. Consider hyphenated numbers (like twenty-fifth) as one word.
   She won first prize.
   They celebrated their twenty-fifth wedding anniversary.

BUT

   The company will have its 145th anniversary soon.

V. PUNCTUATION

Colons

1. The colon is used after the salutation in a business letter (Dear Ms. Cortes:) and after expressions such as *the following, as follows,* and *listed below.* The colon also introduces long quotations.

2. Note that the first word following a colon is capitalized (1) if that word begins a complete sentence or (2) if that word begins on a new line (as, for example, in lists).

   All of us agree with John: Each estimate must be itemized so that we can be sure that the costs are accurate.

Commas

The comma is used in certain letter parts:

1. In the *date line*, the day and the year are separated by a comma.

   January 12, 1982

   Within a sentence, use two commas to set off the year when it follows the month and day.

   Let's meet on May 6, 1985, to discuss it.

2. In *inside addresses* and *return addresses*, the city and state names are always separated by a comma.

   Fargo, North Dakota 58102
   Joplin, MO 64801

   Within a sentence, use two commas to set off the name of a state when it follows the name of a city.

   He moved to Joplin, *Missouri*, in 1979.

   Also, use a comma after each part of an address, except between the state and the ZIP code.

   Please write to Mr. John Neel, 532 East Tanner Street, Knoxville, Tennessee 37916, if you need additional information.

3. In the *writer's identification line*, a comma is used to separate the writer's name and title if both are on the same line.

   Harriet A. Trask, Treasurer

   However, no comma separates the two when they are on two lines.

   Harriet A. Trask
   Treasurer

4. In the *complimentary closing*, a comma is used at the end of the line (in standard punctuation).

   Sincerely yours,
   Cordially,

Commas are also used in the bodies of letters, memos, and reports.

5. Commas are used to set off names used in direct address.

   Do you agree, *John*, that we should approve this expense? (John is the person spoken to, the name in direct address.)

6. Commas are used to separate three or more items in a series.

   Kent, Lara, and Carole were asked to revise their estimates.
   We called Mrs. Ibsen, discussed the problem, and asked for her advice.

7. Commas are used to set off words or phrases in apposition.

   Mr. De Palma, *our general manager,* is on vacation.
   Mrs. Xavier, *owner of the property,* has hired a real estate agent.

   The phrases in italics are called *appositives. Our general manager* is another way of saying *Mr. DePalma. Owner of the property* is another way of saying *Mrs. Xavier.* Appositives are always set off with two commas (unless, of course, they appear at the end of a sentence).

8. Commas are used before the conjunction in a compound sentence.

   Anthony wanted to attend the convention, *but* his manager limited the number of people who could go.
   Bertha did most of the research for the report, *and* she also wrote the first draft.

9. Commas are used after introductory words, phrases, and clauses.
   Yes, the new price is $9.95.
   Incidentally, Mrs. Warren was promoted to regional manager last month.
   To be eligible for the rebate, customers must show proof of purchase.
   When Ms. Silver arrives, please give her this package.

10. Commas are used to separate adjectives that modify the same noun.

    They developed a *bright, interesting* brochure to send to all customers.
    Mr. Jonas always writes *long, rambling, wordy* reports.

11. Commas are used to separate parenthetical comments from the rest of the sentence.

    In my opinion, we should conduct a marketing survey.
    We should, *in my opinion*, conduct a marketing survey.

Hyphens

1. The hyphen is most commonly used to join words—or parts of words—together. Use hyphens for all words and compound words that are hyphenated in your dictionary. In addition, use hyphens for temporary compounds—adjectives joined together to modify a noun.

   He needs a *3-inch* pipe to fix this. (*3 inches* of pipe)

Amy has an office in a *50-story* building. (*50 stories* high)

We bought a *low-intensity* lamp. (A lamp of *low intensity*)

2. Use hyphens for fractions (two-thirds, one-half) and for numbers such as twenty-one and twenty-first.

3. Also, use hyphens to show end-of-line word breaks, but follow the dictionary to make sure the breaks are correct.

Periods

1. Use a period to end most sentences, including polite requests.

   Will you please send me your check by Monday. (Not really a question)

2. Periods are needed in many abbreviations, but not all. (Check a comprehensive reference manual to be sure.) Be especially careful of spacing with periods in abbreviations: Ph.D., a.m., and so on.

   Note that customary and metric abbreviations for weights and measurements have no periods.

   12 g   14 mm   156.92 k
   9 ft   22 gal   8 in

   The size of each sample should be precisely as described in the specifications: 2.5 cm by 3.6 cm by 3.9 cm.

3. Two-letter state abbreviations also have no periods.

   IL, KS, MS, PA, WY

Semicolons

1. The most common use of the semicolon is to join two closely related independent clauses. Usually, the second clause begins with (or includes) a word such as *therefore, however, consequently,* or *moreover.*

   He mailed the package last Friday; therefore, we should surely receive it by tomorrow.

   Mrs. Anson refused to increase our expense budget; she suggested, moreover, that we decrease it by 10 percent.

   Note the pattern of comma use with *therefore* and *moreover.*

2. In very closely related independent clauses, a semicolon without a connecting word can be used to join the clauses.

   Thursday is Will's birthday; Friday is Mark's.

3. The semicolon replaces the comma in a series when one or more of the items in the series already includes a comma.

   She will be traveling to Kansas City, Kansas; Austin, Texas; Charlotte, North Carolina; and Gardiner, Massachusetts.

Underscores

1. Words used as words are usually underscored.

   Throughout this report, the word underline(receive) is misspelled.

   The term underline(psychosomatic) has an interesting derivation.

2. Titles of books, plays, and other complete works are underscored. (They may also be typed in all capital letters.)

   Three of Mary Stewart's best books are underline(The Crystal Cave), underline(The Hollow Hills), and underline(The Last Enchantment).

VI. SPELLING

When in doubt about spelling, check a dictionary or wordbook. Use a dictionary or a reference manual to check similar words that are spelled and used differently.

accept and *except*
affect and *effect*
there, their, and *they're*
to, too, and *two*

EI and IE

Usually, *ie* is used when the sound is *e*, and *ei* is used when the sound is *a* or when preceded by the letter *c*.

ie: piece, relieve, niece
ei: weight, eighth, neighbor, deceive, receipt, ceiling

Silent E

1. A silent *e* is usually dropped when a suffix (word ending) beginning with a vowel is added to the word.

   believe + ing = believing
   continue + ance = continuance
   disclose + ure = disclosure
   use + able = usable

2. Words ending in *ee* do not drop an *e*.

   agree + able = agreeable
   see + ing = seeing

3. A silent *e* is usually retained when a suffix beginning with a consonant is added to the word. A consonant is any letter other than *a, e, i, o,* and *u*.

   achieve + ment = achievement
   trouble + some = troublesome
   late + ness = lateness
   leisure + ly = leisurely

Exceptions

   acknowledge + ment = acknowledgment
   judge + ment = judgment

Words Ending in IE or Y

1. For words ending in *ie*, change the *ie* to *y* before adding *ing*.

   lie + ing = lying
   tie + ing = tying

2. For words ending in a consonant plus *y*, change the *y* to *i* before adding a suffix.

   heavy + er = heavier
   involuntary + ly = involuntarily

   If the suffix begins with an *i*, do *not* double the *i*.

   library + ian = librarian

3. For words ending in a vowel plus *y*, keep the *y* and add the suffix.

   play + er = player   say + ing = saying

Doubling the Final Consonant

1. A final consonant is *not* doubled when the suffix also begins with a consonant.

   cup + ful = cupful
   kind + ly = kindly

2. One syllable words: A final consonant is doubled when a vowel comes before the consonant and a vowel begins the suffix.

   cup + ed = cupped
   plan + ed = planned
   slim + est = slimmest

3. More than one syllable: A final consonant is doubled before a suffix beginning with a vowel *if the last syllable is accented.*

refer + ed = referred   (accent on *fer*)
begin + ing = beginning   (accent on *gin*)
differ + ed = differed   (accent on *dif*)
total + ing = totaling   (accent on *to*)

Exceptions
Two words that have been coined by computer and word processing users do not fit these rules.

format + ing = formatting
program + ed = programmed

Special Problems
1. Beware of word endings:

able or *ible*: dependable, likeable; forcible, flexible
ant or *ent*: defendant, resistant; dependent, persistent
ance or *ence*: assistance, relevance; intelligence, occurrence
ise or *ize* or *yze*: televise, advertise; summarize, criticize (Analyze is the only commonly used word ending in *yze*.)
cede or *ceed* or *sede*: accede, concede, precede (All words with the "seed" ending except exceed, proceed, succeed, and supersede use the *cede* ending.)

Plurals
1. Most plurals are formed by adding *s* to the singular forms.

report + s = reports
office + s = offices

2. When the singular ends in *s, x, z, ch,* or *sh*, add *es*.

boss + es = bosses
box + es = boxes
church + es = churches
bush + es = bushes
Schwartz + es = Schwartzes
Adams + es = Adamses

3. When the singular word ends in a vowel plus *y*, add *s*.

boy + s = boys
bay + s = bays

4. When the singular word ends in a consonant plus *y*, change the *y* to *i* and add *es*.

company + ies = companies
ply + ies = plies

5. To form the plural of a hyphenated or spaced compound, make the main word plural.

mother-in-law + s = mothers-in-law

6. Some plurals are irregular,

man = men
woman = women
foot = feet

Possessives
1. For most singular words, form the possessive by adding an apostrophe (') plus *s*.

manager/manager's   mother/mother's
boss/boss's   Jones/Jones's   friend/friend's

Notice the distinction in the use of *'s* in the following:

Mark's and Connie's homes (Separate ownership; each noun has an *'s*)
Mark and Connie's homes (Joint ownership; only one noun has an *'s*)

2. If a singular proper name ending in *s* would be hard to pronounce with an apostrophe plus *s*, then add only the apostrophe.

Hastings/Hastings'

3. Plural nouns ending in *s* take only an apostrophe to form their possessives.

managers/managers'   bosses/bosses'
friends/friends'

4. Plural nouns that do not end in *s* take an apostrophe plus *s*.

men/men's   women/women's
children/children's

5. If a compound noun is singular, add an apostrophe plus *s*.

editor in chief/editor in chief's

6. If a compound noun is a plural ending in *s*, then add only an apostrophe.

brigadier generals/brigadier generals'

7. If a compound noun is a plural that does not end in *s*, add an apostrophe plus *s*.

sisters-in-law/sisters-in-law's
runners-up/runners-up's

8. Always use a possessive form before gerunds (*ing* nouns).

Mrs. Jonas commented on *his* arriving late to work.

VII. WORD DIVISION
1. Avoid dividing words.

2. Avoid dividing the first line and the last full line of a paragraph.

3. Avoid dividing the root word in a word with a prefix or suffix. Dividing after a prefix or before a suffix is preferable.

4. Divide words between syllables.

5. When in doubt about correct syllabication, check a dictionary or wordbook.

6. Leave at least two letters on the line with the hyphen, and carry over at least three letters (or two letters plus a punctuation mark).

7. Divide compound words where the two words are joined.

8. Divide hyphenated words at the hyphen.

9. Divide between double consonants (*ship-ping*) unless the double consonants end the root word (*call-ing* or *bless-ing*).

10. Divide after, not before, a single-vowel syllable (*prodi-gal*, not *prod-igal*). If there are two single-vowel syllables together, the words should be divided between these two syllables (*cre-ative*, not *crea-tive*).

11. Do not divide one-syllable words.

12. Do not divide words with six or fewer letters.

13. Do not divide abbreviations, contractions, or figures.

14. Do not divide the last word on a page.

15. Do not divide a word on three consecutive lines. Dividing on two consecutive lines is acceptable.

VIII. REVISION SYMBOLS
CAPITALIZATION

Revision	Edited Draft	Final Copy
Capitalize a letter	texas	Texas
Lowercase a letter	This	this
All capitals	Cobol	COBOL
Lowercase a word	PROGRAM	program
Initial capital only	PROGRAM	Program

CHANGES AND TRANSPOSITIONS

Revision	Edited Draft	Final Copy
Change a word	price is only $10.98 / $12.99 . . .	price is only $12.99 . . .
Change a letter	deductible	deductible
Stet (do not make the change)	price are is only $10.98 . . .	price is only $10.98 . . .
Spell out	on Washburn (Rd.) in . . .	on Washburn Road in . . .
	(2) pens and (4) pencils . . .	two pens and four pencils . . .
Move as shown	. . . on May 1 (write him) . . .	. . . write him on May 1 . . .
Transpose letters	hte	the
Transpose words	most the of staff	most of the staff

DELETIONS

Revision	Edited Draft	Final Copy
Delete a letter and close up	stroke or stroke	stroke
Delete a word	wrote two two checks	wrote two checks
Delete punctuation	report was up to date.	report was up to date.
Delete one space*	good # day #	good day
Delete space	see ing	seeing

* Use marginal notes for clarification.

INSERTIONS

Revision	Edited Draft	Final Copy
Insert a letter	may leave erly	may leave early
Insert a word	in office	in the office
Insert a comma	may leave early . . .	may leave early, . . .
Insert a period	Dr Maria Rodriguez	Dr. Maria Rodriguez
Insert an apostrophe	all the boys hats	all the boys' hats
Insert quotation marks	Move on, she said.	"Move on," she said.
Insert hyphens	up to date report	up-to-date report
Insert a dash	They were surprised even shocked!	They were surprised—even shocked!
Insert parentheses	pay fifty dollars ($50)	pay fifty dollars ($50)
Insert one space	mayleave	may leave
Insert two spaces*	1.The new machine	1. The new machine

* Use marginal notes for clarification.

BLOCK SYMBOLS

Revision	Edited Draft	Final Copy
Identify block	and the catalog will be mailed.	
Insert identified block	Your order will be mailed.	Your order and the catalog will be mailed.
Delete identified block	Your order and the catalog will be mailed.	Your order will be mailed.
Move identified block*	Your order and the catalog will be shipped. The invoice will be mailed.	Your order will be shipped. The invoice and the catalog will be mailed.
Query identified block*	Ed will retire at the age of 96. (Are the numbers transposed? Verify age.)	Ed will retire at the age of 69.
Query identified block*	Make my reservation for June 31. (June has only 30 days. Verify date.)	Make my reservation for June 30.
Query conflicting blocks*	Call me Monday morning at 8 p.m. (*Morning* or *p.m.*? Verify time.)	Call me Monday morning at 8 a.m.

* Use marginal notes for clarification.

GLOBAL SYMBOLS

Revision	Edited Draft		Final Copy
Global insert* (Inserts *red* with every occurrence of *labels*)	red labels all ~~labels~~. The large labels are . . .	⟨∧⟩	all red labels. The large red labels are . . .
Global delete* (Deletes every occurrence of *very*)	a ~~very~~ high profit on a very good item	⟨∂⟩	a high profit on a good item
Global change* (Changes every occurrence of *very high* to *high*)	high a ~~very high~~ profit on a very good item. Our very high profit . . .	⟨—⟩	a high profit on a very good item. Our high profit . . .

* Use marginal notes for clarification. Marginal symbols are circled to indicate changes are global.

FORMAT SYMBOLS: BOLDFACE AND UNDERSCORE

Revision	Edited Draft	Final Copy
Print boldface	Bulletin	**Bulletin**
No boldface	**Bulletin**	Bulletin
Underscore	Title	<u>Title</u>
No underscore	<u>Title</u>	Title

FORMAT SYMBOLS: PAGE AND PARAGRAPH

Revision	Edited Draft	Final Copy
Begin a new page	. . . order was delivered today by *pg.* common carrier. We have all the materials for the product.	. . . order was delivered today by Page 2 common carrier. We have all the materials for the product.
Begin new paragraph	. . . order was delivered today by common carrier. ¶ We have all the materials for the product. It should . . .	. . . order was delivered today by common carrier. We have all the materials for the product. It should . . .
No new paragraph (run-in)	. . . order was delivered today by common carrier. *No* ¶ We have all the materials for the product. It should . . .	. . . order was delivered today by common carrier. We have all the materials for the product. It should . . .
Indent five spaces	<u>5</u> We have the raw materials in our warehouse. Production will . . .	We have the raw materials in our warehouse. Production will . . .

FORMAT SYMBOLS: CENTERING

Revision	Edited Draft	Final Copy
Center line horizontally	⌐TITLE ⌐	TITLE
Center identified block horizontally and vertically*	A MENU Juice and Coffee Scrambled Eggs Toast and Jam	MENU Juice and Coffee Scrambled Eggs Toast and Jam

* Use marginal notes for clarification.

FORMAT SYMBOLS: ALIGNMENT

Revision	Edited Draft	Final Copy
Align horizontally	Coleen answered the letter.	Coleen answered the letter.
Align vertically	Coleen answered the letter today. You should receive it Monday.	Coleen answered the letter today. You should receive it Monday.
	$122.30 22.40 $144.70	$122.30 22.40 $144.70

FORMAT SYMBOLS: SPACING

Revision	Edited Draft	Final Copy
Single-space	ss [xxxxxxxxxx xxxxxxxxxx	xxxxxxxxxx xxxxxxxxxx
Double-space	ds [xxxxxxxxxx xxxxxxxxxx	xxxxxxxxxx xxxxxxxxxx
Triple-space	ts [xxxxxxxxxx xxxxxxxxxx	xxxxxxxxxx xxxxxxxxxx